Margaret Baker

Martin Rowland

Hodder & Stoughton

A MEMBER OF THE HODDER HEADLINE GROUP

Photo credits and Acknowledgements
Thanks are due to the following copyright holders for permission to reproduce photographs:
Fig. 2.4 Stephen Dalton/NHPA; Fig. 3.1a Richard Hamilton Smith/Corbis; Fig. 3.1b Richard Bickel/Corbis; Fig. 3.3 Stephen Dalton/NHPA; Fig. 4.1 Martin Harvey/Corbis; Fig. 4.2a Chris Hellier/Corbis; Fig. 4.2b Dr Jeremy Burgess/SPL; Fig. 5.8a Dr Jeremy Burgess/SPL; page 34 top M. Kalab, Custom Medical Stock Photo/SPL; page 34 middle Simon Fraser/SPL; page 34 bottom Matt Meadows/SPL; page 35 top Darrell Gulin/Corbis; page 35 bottom Roger Hosking, Frank Lane Picture Agency/Corbis.

Every effort has been made to contact copyright holders but if any have been inadvertently overlooked the publishers will be pleased to make the necessary arrangements at the earliest opportunity.

AGA(NEAB)/AQA examination questions are reproduced by permission of the Assessment and Qualifications Alliance.

Orders: please contact Bookpoint Ltd, 130 Milton Park, Abingdon, Oxon OX14 4SB. Telephone: (44) 01235 827720. Fax: (44) 01235 400454. Lines are open from 9.00–6.00, Monday to Saturday, with a 24 hour message answering service. Email address: orders@bookpoint.co.uk.

British Library Cataloguing in Publication Data
A catalogue record for this title is available from the British Library

ISBN 0 340 813563

First published 2004
Impression number 10 9 8 7 6 5 4 3 2 1
Year 2010 2009 2008 2007 2006 2005 2004

Illustrations by Art Construction
Typeset in 10pt Goudy by Tech-Set Ltd
Printed in Spain for Hodder & Stoughton Educational, a division of Hodder Headline, 338 Euston Road, London NW1 3BH

Papers used in this book are natural and recyclable products. They are made from wood grown in sustainable forests. The logging and manufacturing processes conform to the environmental regulations of the country of origin.

Contents

Introduction

About this book

This is not a traditional textbook and will not substitute for one. Instead, it should be used as a revision aid, to help consolidate your recall and understanding and to develop examination skills. The chapters have been written to a common style and they share the following features.

CHAPTER OPENING

Each chapter opens with a summary of what you should know or be able to do. You can use this is a quick reminder of each topic, to reinforce what you know and can do or to prompt you to do further revision.

THE TEXT

Most biology teachers extend the subject beyond the specification content to help you put information in a broader context or to interest you and motivate your learning. We have not done this: this book contains only what is in Specification A of the Assessment and Qualifications Alliance (AQA) and nothing else. If you look at questions in past examination papers that test recall, you will find the answers are in this book. If examination questions go beyond the content of this book, the examiner will have given you the additional information you need within the question. Do not be put off by long examination questions: they usually indicate that skills other than recall are being tested. Questions testing recall are usually very short.

We have kept the amount of continuous prose to a minimum. Instead, we have presented information in ways that we know students find easy and quick to use. These include lists of bullet points, tables and annotated diagrams. Not only will this help you to structure your revision of facts, processes and principles but, by using the tables and diagrams, it will help you to develop interpretive and analytical skills that are tested in examinations. Wherever possible, we have used diagrams that you could reproduce yourself in an examination.

EXAMINER'S TIPS

Between us, we have considerable experience of setting and marking A2 examination papers. Using this experience, we have included tips in each chapter that will help you to improve your examination performance. In these tips, we point out errors committed by past candidates or inform you of good practice.

IN-TEXT QUESTIONS

Examinations test skills, so you need to develop these as you learn and revise. Each chapter has in-text questions that encourage you to become actively involved in your learning. Reading the text without interacting with what you have just read is not an effective way to learn or revise.

Answers have been provided for all the in-text questions at the back of the book. Try to answer the questions yourself and use our answers only to check your own. Remember, you must focus on developing the skills that will be tested in your examinations.

EXAMINATION QUESTIONS

Each chapter contains questions from past A2 examinations that are reproduced or adapted by permission of the Assessment and Qualifications Alliance (AQA).

One question in each chapter contains a student's answer that has been marked by us. We have also given a commentary to explain why the candidate's answer gained, or failed to gain, marks. The more you learn to think like an examiner, the better your grade will be.

Good luck – we look forward to marking your scripts.

1 Transmission of genetic information

After revising this topic, you should be able to:

- describe the behaviour of chromosomes during **meiosis**
- explain the importance of meiosis in **maintaining the chromosome number** following fertilisation and in contributing to **genetic variation**
- list the main differences between **meiosis** and **mitosis**
- understand the meaning of the basic terms used in genetics
- solve genetic problems showing:
  - monohybrid inheritance
  - codominance
  - multiple alleles
  - dihybrid inheritance
  - sex linkage
- calculate and interpret the **chi-squared** test

Meiosis

- Meiosis is a process of nuclear division in which the number of chromosomes is halved.
- It is usually associated with sexual reproduction.
- Meiosis produces four haploid cells.
- The cells produced are usually sex cells (gametes).
- The cells show genetic variation.
- Variation comes from two processes in meiosis: crossing over and independent assortment.

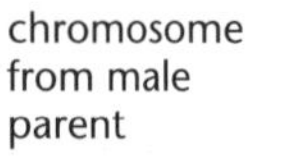

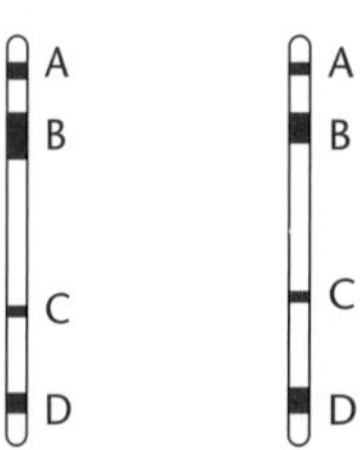

Letters A-D represent four different genes. Each chromosome could contain different alleles of each gene.

FIGURE 1.1 Homologous pairs.

REMEMBER

In each body cell:

- One set of chromosomes (**haploid**) is provided by the **female parent**.
- One set of chromosomes (**haploid**) is provided by the **male parent**.
- Each set of chromosomes carries the **same genes** in the **same sequence**.
- The chromosomes may **not be identical**, as they may **not carry the same alleles** of each gene.

The cells that are undergoing meiosis have two sets of chromosomes (diploid) in **homologous pairs**.

Meiosis involves two separate stages, referred to as **meiosis I** and **meiosis II**.

SUMMARY OF MEIOSIS

During the cell cycle the DNA in each chromosome will replicate identically. So at the beginning of meiosis, each chromosome is a double structure consisting of a pair of chromatids.

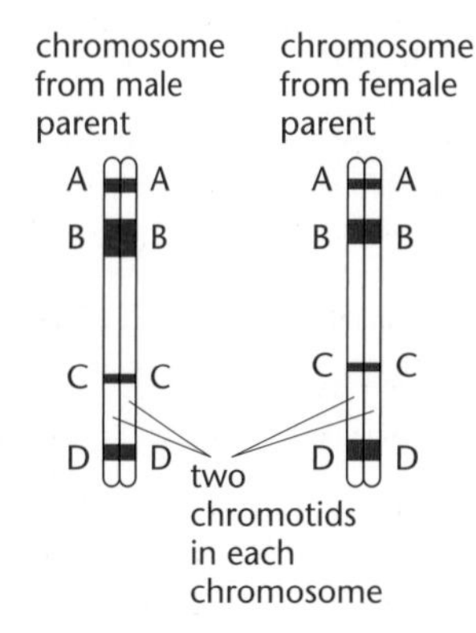

FIGURE 1.2 Chromatids.

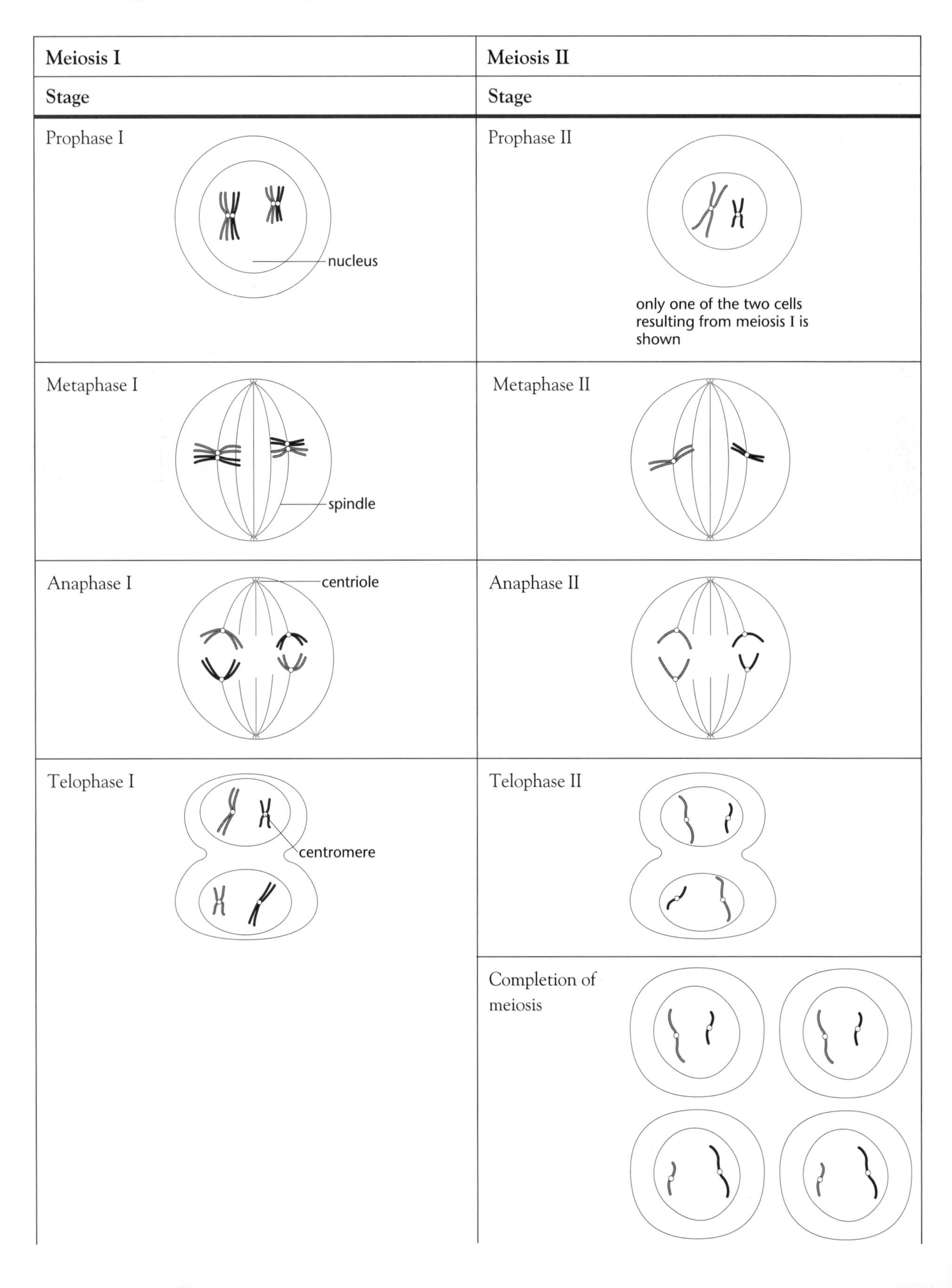

Meiosis I	Meiosis II
Stage	**Stage**
Prophase I	Prophase II
Metaphase I	Metaphase II
Anaphase I	Anaphase II
Telophase I	Telophase II
	Completion of meiosis

OVERALL	OVERALL
◆ Chromosomes come together in their **homologous pairs**. ◆ Each homologous pair appears as a **group of four chromatids (a bivalent)**. ◆ Each pair of chromosomes pulled apart. ◆ They move to opposite ends of the cell: to **opposite poles of the spindle**. ◆ This results in **one complete set of chromosomes** at each pole of the spindle.	◆ Each chromosome consists of **two chromatids**. ◆ Each pair of chromatids is pulled apart. ◆ Once separated, each chromatid is known as a **chromosome**. ◆ This results in one chromosome **from each homologous pair** in each cell.

- A chromosome may be a single strand of DNA, or it could consist of two strands following replication.
- If a chromosome consists of two strands, each strand is called a chromatid.

1 During what phase of the cell cycle does DNA replication take place?

Genetic variation

There are two events that give rise to genetic variation:

- Crossing over
- Independent assortment

Event	Stage	Description
Crossing over	Prophase I	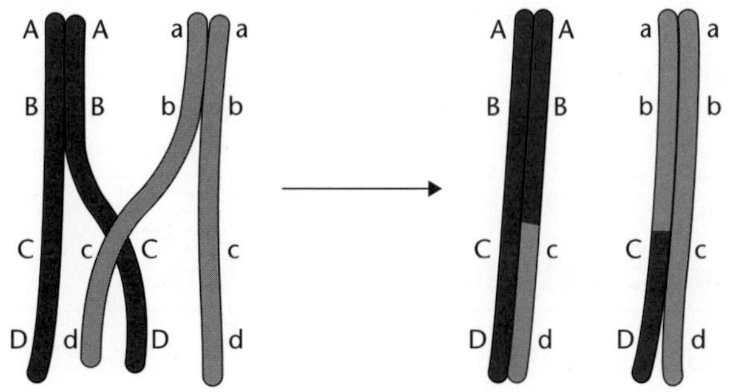 A A a a B B b b C c C c D d D d → A A a a B B b b C c C c D d D d ◆ When homologous chromosomes associate, the four chromatids **twist** around each other. ◆ During this process, two chromatids may **break and rejoin**. Chiasmata is the place where chromatids are seen to cross over each other. ◆ If one of the chromatids involved comes from the male parent and the other from the female parent, there would be a **new mixture** of genetic material.

Independent assortment	Metaphase I	◆ The bivalents line up on the equator **randomly**. ◆ For each bivalent, the chromosome from the female parent may line up on the 'north' **or** 'south' side of the equator. ◆ The same is true for the homologous chromosome from the male parent. ◆ This happens randomly for all bivalents in the cell. ◆ When these bivalents separate, depending on how they have lined up on the equator, **different combinations of chromosomes** will end up at the **poles**. ◆ This will result in daughter cells at the end of meiosis I **with different mixtures of maternal and paternal chromosomes**. equator of spindle chromosome from female parent chromosome from male parent

Meiosis and Mitosis

There are a number of basic differences between these two types of cell division.

Meiosis	Mitosis
The number of chromosomes present in each daughter cell is **halved**.	The number of chromosomes present in each daughter cell **remains the same**.
Meiosis involves **two divisions** of the nucleus, and results in **four daughter nuclei** being formed.	Mitosis involves one division of the nucleus and results in **two daughter cells** being formed.
During meiosis I, chromosomes come together in their **homologous pairs**, each pair appearing as a group of **four chromatids**: a bivalent.	During mitosis, the chromosomes do **not** come together in homologous pairs. Each chromosome consists of **two chromatids**.
Crossing over may result in genetic variation.	Crossing over may take place, but is of **identical genetic material**: so it never results in genetic variation.

Genetics and genetic problems

It is important that you know some of the basic terms used in genetics.

Term	Definition
Gene	◆ A length of DNA at a particular location (locus) on a chromosome that codes for a particular protein and determines a particular feature. ◆ For example there is a gene for rolling the tongue.

Allele	◆ A particular form of a given gene. ◆ The tongue-rolling gene has an allele for the ability to roll the tongue and an allele for the inability to roll the tongue.
Homozygous	The cells in an organism have identical alleles for a gene.
Heterozygous	The cells in an organism have different alleles for a gene.
Dominant	◆ A dominant allele is one that, if present, will always be expressed in the phenotype. Usually it is represented by an upper case letter. ◆ The ability to roll the tongue is the result of a dominant allele and is represented by R.
Recessive	◆ A recessive allele is only expressed when cells are homozygous, i.e. both alleles are identical. ◆ It is usually represented by a lower case letter. ◆ The allele for the inability to roll the tongue is recessive and is represented by r.
Genotype	This describes the alleles that are present, e.g. RR or Rr or rr.
Phenotype	This is the expression of the genotype: an observable or measurable characteristic, e.g. ◆ RR – results in the ability to roll the tongue ◆ Rr – results in the ability to roll the tongue ◆ rr – results in the inability to roll the tongue

? 2 DNA is a polymer. Name the monomer of which it is made.

Monohybrid inheritance with dominance

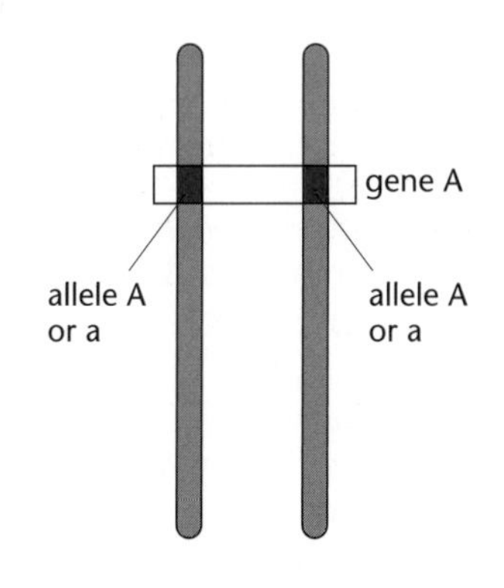

FIGURE 1.3 Homologous chromosomes, for gene A.

- Monohybrid inheritance occurs when **one** phenotypic characteristic is controlled by **one** gene.
- The gene can have **two alleles**, A or a.
- In these examples, one allele is **dominant** to another.

EXAMINER'S TIP: It doesn't matter how simple the question is, always lay out the problem systematically.

? 3 Name a phenotypic characteristic that may be advantageous for: a) a guard dog; b) a flower; c) a potential marriage partner.

Let's do a simple question to show the technique.

Question

In pea plants, the allele for tall stems is dominant to the allele for short stems. If two tall-stemmed plants (both heterozygous) are crossed, what percentage of the offspring will be short-stemmed plants?

Answer

Parental phenotype	Tall	Tall
Parental genotype	Tt	Tt
Parental gametes	(T) or (t)	(T) or (t)

	(T)	(t)
(T)	TT	Tt
(t)	Tt	tt

Offspring genotype	TT	Tt	Tt	tt
Offspring phenotype	Tall	Tall	Tall	Short

The answer to the question is clear: $\frac{1}{4}$ or 25% will be short-stemmed plants.

Remember the following ratios when dealing with monohybrid inheritance with dominance.

Ratio	Cross
3 : 1	Two heterozygous parents
1 : 1	◆ Heterozygous parent crossed with a homozygous recessive parent ◆ This is also known as a **test cross**

EXAMINER'S TIP

- Always write your gametes in circles as it clearly distinguishes them from genotypes. For example, can you see the difference between Tt and T t? An examiner may not and, therefore, you might not be given credit. If you are reluctant to use circles, then separate the gametes clearly by using a comma between them, e.g. T, t.
- Never rush even the most simple genetics question: marks are always given for steps in the procedure. If you only give the answer, you can only get one mark – and more may be on offer.
- If you do spider diagrams, mistakes will occur and it is sometimes difficult to go back and check.
- If you are given letters to use to represent the alleles, use them.
- If you are not given letters to use, be sensible in your choice. Traditionally, the dominant allele is given the capital letter, but if you choose the letters C, L, O, P or S (and, depending on your style of writing, this could include more) the capital letter is very similar to the small letter and, in the stress of an exam, you might write it unclearly or read it incorrectly.

EXAMINER'S TIP

Regardless of the simplicity of the genetics, always draw a contingency table. This prevents confusion and ensures that each possible combination of alleles from the two parents is considered.

?

4 If each of the following pairs of characteristics is controlled by a gene with two alleles (one dominant and one recessive) suggest letters you would use to represent them in a genetic diagram:
a) Red flower dominant to white flower
b) Smooth seed pod dominant to wrinkled seed pod
c) Green feathers dominant to yellow feathers.

Monohybrid inheritance with codominance

- Monohybrid inheritance occurs when **one** characteristic is controlled by **one** gene.
- Each gene can have **two alleles**.
- In these examples, one allele is **not dominant** to another – both show their effect in the phenotype
- It is not possible, therefore, to represent these alleles by upper or lower case letters. In these examples, we give the gene an upper case letter, e.g. C, and use a **superscript** to represent the allele, thus, C^R or C^W.

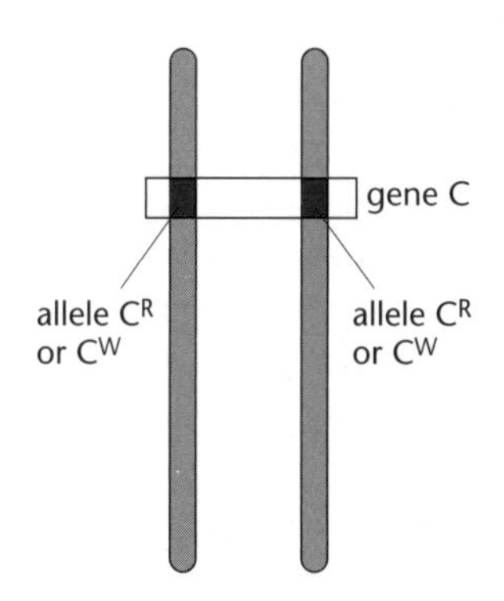

FIGURE 1.4 Homologous chromosomes, for gene C.

Question

In roan cattle, a pair of codominant alleles control the colour of the hair. When a red-haired bull was mated with a white-haired cow, the offspring had roan-coloured hair, i.e. a coat with both red and white hairs. If a roan bull were mated with a roan cow what is the probability that they will have a white calf?

Answer

Parental phenotype	Roan		Roan	
Parental genotype	C^WC^R		C^WC^R	
Parental gametes	(C^W) or (C^R)		(C^W) or (C^R)	

	(C^W)	(C^R)
(C^W)	C^WC^W	C^WC^R
(C^R)	C^WC^R	C^RC^R

Offspring genotype	C^WC^W	C^WC^R	C^WC^R	C^RC^R
Offspring phenotype	White	Roan	Roan	Red

The answer to the question is clear. There is a $\frac{1}{4}$ or 25% probability of the calf being white.

Remember these ratios when dealing with monohybrid inheritance with codominance:

Ratio	Cross
1 : 2 : 1	Two heterozygous parents.
1 : 1	Heterozygous parent crossed with homozygous recessive parent. Also known as a test cross.

EXAMINER'S TIP

The give-away which identifies whether a characteristic is controlled by codominant alleles is that a cross between two different parental phenotypes produces an offspring with a third phenotype, i.e. one that is different to that of either parent.

Monohybrid cross with multiple alleles

- Once again, with a monohybrid cross, **one characteristic** is controlled by **one gene**.
- Each gene can have **more than two alleles**. (Remember an organism can only have two – one inherited from each parent.)
- Because there is more than one allele to be represented, once again upper and lower case letters are not enough.
- Once again, we give the gene a letter and use a **superscript** to represent the allele. For example, I^O, I^A and I^B represent the three alleles controlling blood groups:
  - I^A is dominant to I^O
  - I^B is dominant to I^O
  - I^A is codominant to I^B.

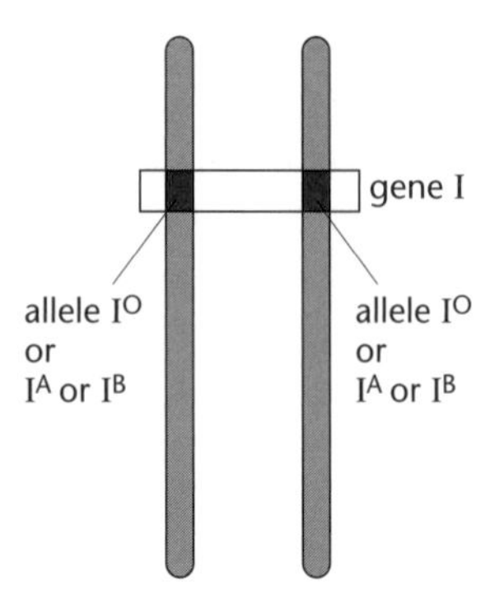

FIGURE 1.5 Homologous chromosomes, for gene I.

Question

A mother with blood group A, whose father was blood group O, marries a man with blood group AB. What is the probability that their first child will be a girl with blood group B?

Answer

Parental phenotype	Blood group A	Blood group AB
Parental genotype	I^AI^O We know this because her father must have given her an I^O allele as he was blood group O genotype I^OI^O	I^AI^B This is the only possible genotype for a person with blood group AB
Parental gametes	(I^A) (I^O)	(I^A) (I^B)

	(I^A)	(I^B)
(I^A)	I^AI^A	I^AI^B
(I^O)	I^AI^O	I^BI^O

Offspring genotype	I^AI^A	I^AI^B	I^AI^O	I^OI^B
Offspring phenotype	Blood group A	Blood group AB	Blood group A	Blood group B

The answer to the question involves two steps:

- The probability of being blood group B is $\frac{1}{4}$.
- The probability of being a girl is $\frac{1}{2}$.
- To arrive at the answer you have to multiply these two probabilities: $\frac{1}{2} \times \frac{1}{4} = \frac{1}{8}$.

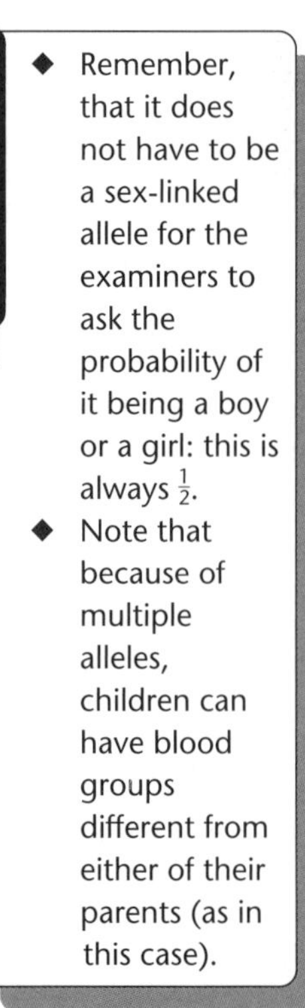

- Remember, that it does not have to be a sex-linked allele for the examiners to ask the probability of it being a boy or a girl: this is always $\frac{1}{2}$.
- Note that because of multiple alleles, children can have blood groups different from either of their parents (as in this case).

? **5 If both parents have blood group O, what is the probability of their next child being a girl with blood group O?**

Dihybrid inheritance with dominance

- Dihybrid inheritance occurs when **two characteristics** are controlled by **two genes** on separate chromosomes.
- Each gene has two alleles. Each is represented by a different letter.
- In these examples, one allele is **dominant** to another, e.g. T to t and F to f.

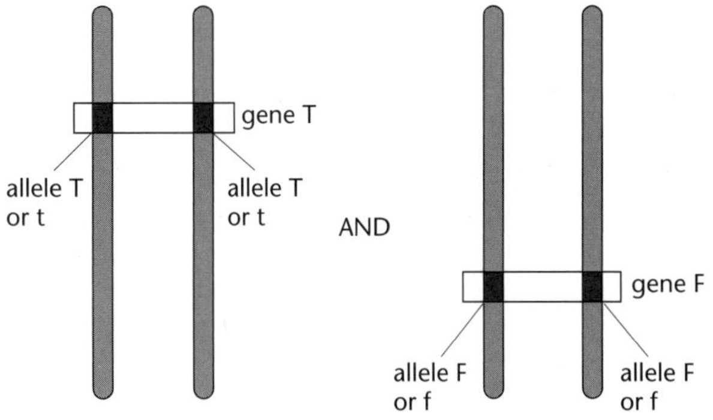

gene T
allele T or t
allele T or t
AND
gene F
allele F or f
allele F or f

Figure 1.6 Homologous chromosomes, for genes T and F.

Question

The allele for tall stem (T) is dominant to that for short stem (t); and the allele for purple flowers (F) is dominant to the allele for white flowers (f). A plant, which is heterozygous for both tall stem and purple flowers, is self-pollinated. What is the probability that the seeds produced will give rise to short-stemmed plants with white flowers?

Answer

Parental phenotype	Tall and purple	Tall and purple
Parental genotype	TtFf	TtFf
Parental gametes	(TF) (Tf) (tF) (tf)	(TF) (Tf) (tF) (tf)

EXAMINER'S TIP

- This type of inheritance is where most candidates make a mistake.
- Gametes are haploid. Each gamete will contain one chromosome from each pair. In this example, each gamete will contain one of the alleles for height and one of the alleles for flower colour.
- Therefore, the gamete must have one of each letter. NEVER two the same, (Tt) or (Ff).
- Because chromosome assortment is random, any allele can go with any allele.

	(TF)	(Tf)	(tF)	(tf)
(TF)	TTFF	TTFf	TtFF	TtFf
(Tf)	TTFf	TTff	TtFf	Ttff
(tF)	TtFF	TtFf	ttFF	ttFf
(tf)	TtFf	Ttff	ttFf	ttff

Offspring genotype	Tall/purple	Tall/white	Short/purple	Short/white
Offspring phenotype	~~IIII~~ IIII = 9	III = 3	III = 3	I = 1

EXAMINER'S TIP

- This again is a step where mistakes occur.
- Take your time and be systematic in your approach.
- Work out the phenotype of each square in turn, cross it out clearly and distinctly (or shade as done here) and tally the results.

The answer to the question is that $\frac{1}{16}$ of the offspring will be short-stemmed with white flowers.

Remember the following ratios when dealing with dihybrid inheritance with dominance:

Ratio	Cross
9 : 3 : 3 : 1	Both parents heterozygous for both genes.
1 : 1 : 1 : 1	◆ A heterozygous parent for both genes crossed with a parent homozygous recessive for both genes. ◆ This is also known as a test cross.

EXAMINER'S TIP: Although a dihybrid cross has two characteristics and the example shown above is of a dominant/recessive relationship, this is not always the case. The alleles could be dominant, recessive, codominant, multiple alleles or sex-linked, so read the question carefully to work out the relationship of the alleles involved.

Sex linkage

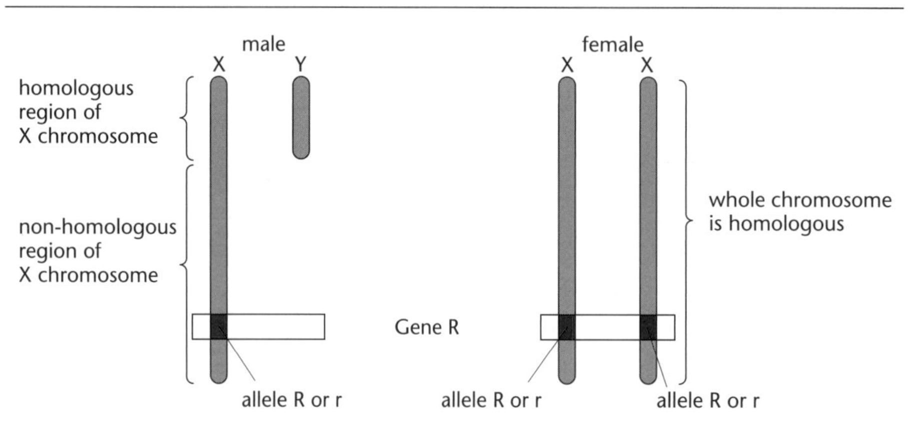

FIGURE 1.7 Chromosomes.

- Sex in humans is determined by **sex chromosomes**: the male has **one X** and **one Y** chromosome (XY); the **female** has **two X** chromosomes (XX).
- The X chromosome is **larger** than the Y chromosome.
- A gene is described as being sex-linked if it is found on **one** of the sex chromosomes.
- There are many genes controlling features on these chromosomes. The most common are red/green colour blindness, haemophilia and Duchennes muscular dystrophy.
- In males, because the Y chromosome is smaller, the X chromosome **does not have a matching section** on the Y chromosome.
- Thus, alleles in this region of the X chromosome do **not** have a **corresponding** allele on the Y chromosome.
- Therefore, **recessive characteristics** may be shown in the phenotype when **only one recessive allele is present** – on the X chromosome.
- The usual way of representing this is as follows. In colour blindness, the dominant allele of this gene, R, codes for the normal pigment, so a person possessing this allele has normal colour vision. The recessive allele, r, codes for the faulty pigment that produces colour blindness. We represent the dominant allele, carried on the X chromosome as X^R and the recessive allele as X^r. The letter X shows us that the allele is on the X chromosome.

Question

A man who is red-green colour-blind, marries a woman who is homozygous for the allele for normal colour vision. Show by means of a genetic diagram, the possible genotypes and phenotypes of their children.

Answer

Parental phenotype	Colour-blind man	Woman with normal vision
Parental genotype	X^rY	X^RX^R
Parental gametes	(X^r) (Y)	(X^R)

EXAMINER'S TIP

- The parental gamete line shows all possible alternative gametes. If there is only one possible gamete, there is no need to write it twice, as with the woman with normal vision: she can only produce (X^R) gametes.
- The Y chromosome does not carry this allele, so the allele for colour blindness does not have to be represented.

	(X^R)
(X^r)	X^RX^r
(Y)	X^RY

Offspring genotype	X^RX^r	X^RY
Offspring phenotype	Girl with normal vision. She is heterozygous; so she is able to pass the X^r allele on to her children. She is, therefore, referred to as a CARRIER.	Boy with normal vision.

EXAMINER'S TIP

- To be female, you must have two X chromosomes. One is donated by your mother the other by your father. Note that a father only has one X chromosome, so if a girl is produced, she must have this X chromosome, whatever allele it is carrying.
- To be male, you have one X and one Y chromosome. The Y chromosome must come from your father, and therefore, the X chromosome must come from your mother. So males with sex-linked conditions inherit the sex-linked alleles from their mothers.
- These conditions are much more common in males than in females, because females must inherit two chromosomes with the recessive allele for the condition to show in the phenotype, whereas males need only inherit one.
- Sex-linked characteristics are never passed from father to son if the gene is on the X chromosome.

6 Haemophilia is a genetically controlled condition caused by a recessive allele which is sex-linked.

a) Explain how it may be possible for a girl to have haemophilia.

b) Deduce the genotypes of her mother and father.

Pedigrees

These are ways of representing the **inheritance of a genetic condition** over **several generations** in a particular family.

- ● represents a female with the feature
- ○ represents a female without the feature
- ◐ represents a female carrier, but will not be used in these examples
- ■ represents a male with the feature
- □ represents a male without the feature

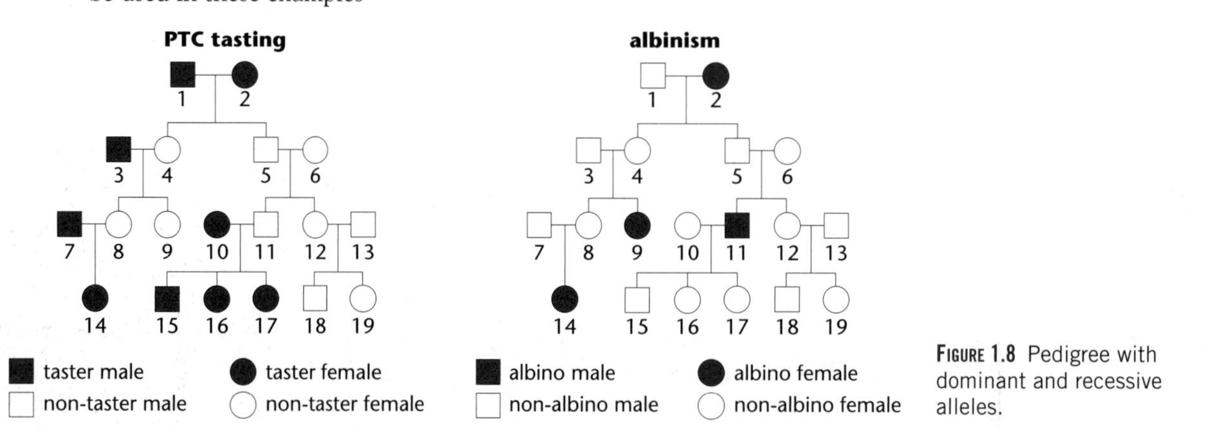

FIGURE 1.8 Pedigree with dominant and recessive alleles.

<table>
<tr><th>Dominant feature</th><th>Recessive feature</th></tr>
<tr><td colspan="2">You must always look out for the individuals showing the recessive condition, as these will be homozygous.</td></tr>
<tr><td>◆ The recessive condition is not being able to taste the chemical PTC. This is represented by ○ or □.
◆ Individuals 4 and 5 are examples.</td><td>◆ The recessive condition means that you are albino. This is represented by ● or ■.
◆ Individuals 9 and 11 are examples.</td></tr>
<tr><td colspan="2">In an examination, you may not be told whether the feature is dominant or recessive. You have to work it out.</td></tr>
<tr><td>◆ From the pedigree you can see that individuals 1 and 2 are tasters, but their children 4 and 5 are not.
◆ If this condition was recessive, then both parents would be homozygous.
◆ They would pass on only recessive alleles to all their children.
◆ The children would have the same phenotype and genotype as their parents.
◆ Clearly this is not the case, therefore the condition must be dominant, and both parents must be heterozygous for the condition.</td><td>◆ From this pedigree you can see that individuals 7 and 8 are normally pigmented, but their daughter, 14, is albino.
◆ This situation can only arise if the condition is recessive.</td></tr>
<tr><td colspan="2">Once you have determined whether the condition is dominant or recessive, you can determine the genotype of other family members.</td></tr>
<tr><td>Question What is the genotype of individual 3?
Answer
◆ This individual is a male taster and therefore must either be homozygous or heterozygous, TT or Tt.
◆ He has a daughter, 9, who is a non-taster, therefore homozygous recessive, tt.
◆ Individual 9 must get one of the alleles from her dad.
◆ This makes individual 3 heterozygous, Tt.</td><td>Question What is the genotype of individual 6?
Answer
◆ The individual is a normally pigmented female, therefore she must be homozygous or heterozygous, NN or Nn.
◆ She has a son, 11, who is albino; therefore, homozygous recessive, nn.
◆ Individual 11 must get one of his alleles from his mother. (One allele is inherited from each parent.)
◆ This makes individual 6 heterozygous, Nn.</td></tr>
</table>

- When looking at pedigrees, always look for specific clues in part of the pedigree.
- One of the biggest clues is that the parents have a different phenotype to their children's.
- Never be influenced by numbers. Each genotype is the result of chance, so the fact that there are many individuals in a family with a condition does not always mean that the condition is dominant.

Sex-linked pedigrees

Figure 1.9 represents part of a pedigree of a family that is affected by haemophilia. On **each X chromosome** there is a gene for blood clotting that is **totally absent** from the **Y chromosome**. The gene has two alleles, the dominant allele, H, that results in normal clotting and the recessive allele, h, that results in very slow clotting – haemophilia.

To determine the genotypes of individuals in a pedigree of a sex-linked feature, begin with a genotype of which you can be **certain**. This can only be:

1. an affected male, X^hY, the affected X chromosome coming from his **mother**, or
2. an affected female, X^hX^h, **each parent** passing on an affected X chromosome.

You can work out others from this starting point.

EXAMINER'S TIP

When answering any questions based on a pedigree, write down on the diagram all the genotypes of which you can be certain right from the beginning.

Question

What are the genotypes of individuals 4 and 10 in Figure 1.9?

Answer

- Individual 4 is the mother of individual 9, a boy with haemophilia, therefore X^hY.
  - The X^h chromosome can only have come from his mother, individual 4.
  - Individual 4 is unaffected, and so must be heterozygous, a carrier, X^HX^h.
- Individual 10 is the daughter of individual 6, a man with haemophilia, X^hY.
  - She is unaffected, so could be X^HX^H or X^HX^h.
  - She must inherit two X chromosomes: one from her father, individual 6, and one from her mother.
  - She must inherit X^h from her father and, therefore, her genotype must be X^HX^h.

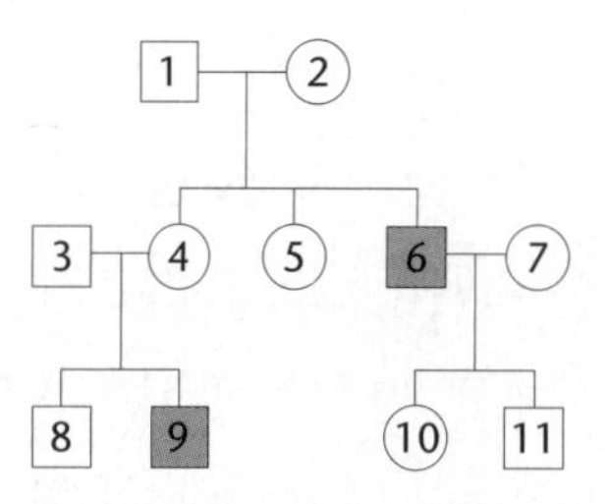

FIGURE 1.9 Sex-linked pedigree.

The chi-squared (χ^2) test

This test allows you to determine the **probability** that what we **observe** is what we would **expect**.

- For example, if we crossed two heterozygous tall plants, we would expect $\frac{1}{4}$ of the offspring to be short. When the test is performed, the actual (observed) results obtained are 80 tall : 20 short. With a hundred plants, we would have expected $\frac{3}{4}$ to be tall and $\frac{1}{4}$ to be short. Therefore, 75 tall : 25 short.
- The chi-squared test allows us to decide whether the difference between the observed and expected values is **due to chance** or **not**.

The formula for the chi-squared test is:

$$\chi^2 = \sum \frac{(O - E)^2}{E}$$

O = observed value
E = expected value
$\sum$ = the sum of

Let's go through this using an example:

Question

Pure breeding *Drosophila* with straight wings and grey bodies were crossed with pure breeding curled-winged, ebony-bodied flies. All the offspring were straight-winged and grey-bodied. Female flies from the offspring were then crossed with curled-winged, ebony-bodied males, giving the following results:

Phenotype	Number
Straight wing, grey body	113
Straight wing, ebony body	30
Curled wing, grey body	29
Curled wing, ebony body	115

a) Give the ratio of the phenotypes expected in a dihybrid cross such as this.
b) Use the χ^2 test to explain these results.

Before you go in to chi squared you must be confident about predicting ratios or do the genetics from scratch.

Answers

a) This is a **dihybrid test cross**, so the expected ratio would be 1 : 1 : 1 : 1.
b) It is again very important to set this out logically. Below are the important steps to follow:

1. Create a **null hypothesis**.
   - This is always that there is **no** difference between the observed and expected values.

2. Work out the **expected** values.
   - Do this by adding all the numbers obtained:
     113 + 30 + 29 + 115 = 287
   - In this case, divide by 4 to give the numbers of each phenotype you would expect.
     – This gives 71.75 as the expected number for each phenotype.
     – Do not round to whole figure e.g. 72.
     – (If the expected ratio was 9 : 3 : 3 : 1, then you would divide by 16 and multiply by 9, then 3, then 3, then 1 to give the expected values.)

3. Create a **table** to calculate chi-squared values.

Phenotypes	Observed number (O)	Expected number (E)	O – E	$(O - E)^2$	$\frac{(O - E)^2}{E}$

This table will fit all situations and all you have to do is put in the specific details for this chi-squared calculation.

Phenotypes	Observed number (O)	Expected number (E)	O – E	$(O - E)^2$	$\frac{(O - E)^2}{E}$
Straight wing, grey body	113	71.75	113 – 71.75 = **+41.25**	+ 41.25^2 = **1701.6**	$\frac{1701.6}{71.75}$ = **23.72**
Straight wing, ebony body	30	71.75	30 – 71.75 = **–41.75**	– 41.75^2 = **1743.1**	$\frac{1743.1}{71.75}$ = **24.29**
Curled wing, grey body	29	71.75	29 – 71.75 = **–41.75**	– 41.75^2 = **1743.1**	$\frac{1743.1}{71.75}$ = **25.47**
Curled wing, ebony body	115	71.75	115 – 71.75 = **+43.25**	– 43.25^2 = **1870.6**	$\frac{1870.6}{71.75}$ = **26.06**

$$\chi^2 = \sum \frac{(O - E)^2}{E} = 23.72 + 24.29 + 25.47 + 26.07$$

Therefore $\boldsymbol{\chi^2 = 99.\ 55}$

4. Use a chi-squared probability table.

To interpret this value you have to use a **chi-squared probability table**, which will show whether the null hypothesis can be **accepted** or **rejected**. These come in a variety of forms, but the principle is always the same. Look at the table below.

Degrees of freedom	Probability, p					
	0.99	0.5	0.1	0.05	0.01	0.001
1	0.00	0.455	2.71	3.84	6.64	10.83
2	0.02	1.386	4.61	5.99	9.21	13.82
3	0.115	2.366	6.25	7.82	11.35	16.27
4	0.297	3.357	7.78	9.49	13.28	18.47
5	0.554	4.351	9.24	11.07	15.09	20.52

5. Calculate the degrees of freedom.

- The number of degrees of freedom is always one less than the number of categories.
- In our example there are 4 categories (possible phenotypes) and therefore:

  4 – 1 = 3

- There are 3 degrees of freedom.
- This means that you need only consider the numbers in that row of the table.

Degrees of freedom	Probability, p					
	0.99	0.5	0.1	0.05	0.01	0.001
1	0.00	0.455	2.71	3.84	6.64	10.83
2	0.02	1.386	4.61	5.99	9.21	13.82
3	0.115	2.366	6.25	7.82	11.35	16.27
4	0.297	3.357	7.78	9.49	13.28	18.47
5	0.554	4.351	9.24	11.07	15.09	20.52

6. Work out the probability of your null hypothesis being accepted.
   - Using the chi-squared value you have calculated, look along the row until you reach a χ^2 value close to the one calculated.
   - In our example, 99.55 is greater than 16.27.
   - This means that the probability that our null hypothesis can be accepted is less than 0.001. This is often represented as $p < 0.001$.

7. Interpret your chi-squared value.

Degrees of freedom	Probability, p					
	0.99	0.5	0.1	0.05	0.01	0.001
1						
2						
3	0.115	2.366	6.25	7.82	11.35	16.27
4						
5						

Below is a simple way to interpret this rather complicated table.

- The table shows the **probability of the null hypothesis being valid**, i.e. that the expected and observed values are the same.
- If these values are **identical** then the χ^2 **value** will be 0.
- The **probability** associated with that value of χ^2 would be 1.
- Theoretically, if the observed and expected values were **very, very different**, the χ^2 value would be **very large** and the probability of the null hypothesis being correct would be 0.
- We have to agree a probability value above which we accept the null hypothesis and below which we reject it.
- This critical value has been agreed as 0.05 by biologists, see the grey shaded area in the table.
- You could also consider the values as a percentage, with 1 being 100%. So the critical value of 0.05 could also be expressed as 5%.

- If your chi-squared value is in the dotted area:
  - **accept** the null hypothesis
  - any **difference** between the observed and the expected values is **due to chance**
  - there is no **significant difference** between the observed and expected results.
- If your chi-square value is in the **red** area or beyond:
  - **reject** the null hypothesis
  - the difference between the observed and expected values is **not due to chance**
  - the difference between the results is **significant**; thus there is likely to be a reason for the difference between the observed and expected values other than chance.

In this example

- The probability <0.001.
- Therefore the null hypothesis is rejected.
- The difference between the observed and expected values/results is not due to chance.
- The difference in the results is significant.
- Our results **cannot** be accepted as corresponding to the expected 1 : 1 : 1 : 1 ratio.

- Do not confuse 0.5 with 0.05.
- 0.5 means 50%, not 5%.
- Rejecting a null hypothesis does not give you a reason why there is a significant difference. You still have to come up with a possible suggestion for it.
- Never write that the null hypothesis is right or wrong. You can only accept or reject a null hypothesis.

WORKED EXAM QUESTIONS

1 Cyanide is a poisonous substance. Cyanogenic clover plants produce cyanide when their tissues are damaged. The ability to produce cyanide is controlled by genes at loci on two different chromosomes. The dominant allele, **A**, of one gene controls the production of an enzyme which converts a precursor to linamarin. The dominant allele, **L**, of the second gene controls the production of an enzyme which converts linamarin to cyanide. This is summarised in the diagram.

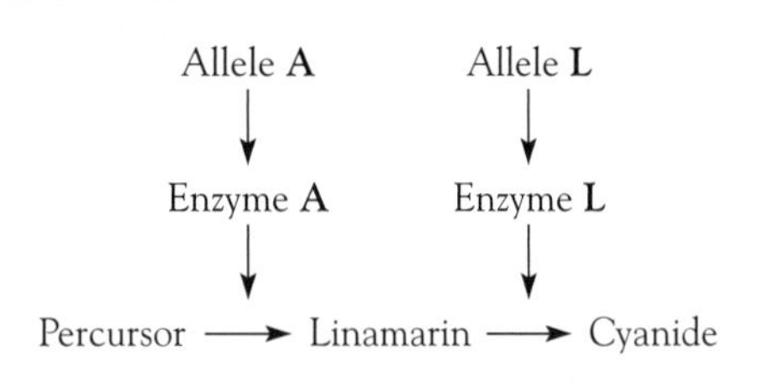

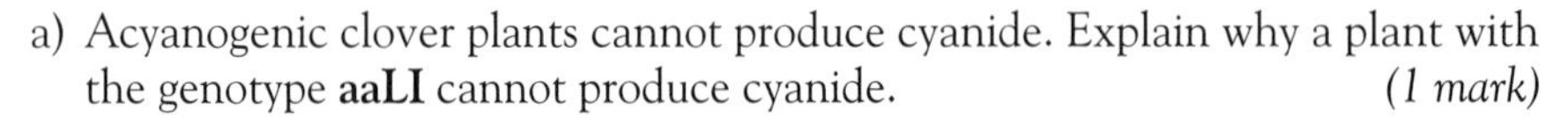

a) Acyanogenic clover plants cannot produce cyanide. Explain why a plant with the genotype **aaLl** cannot produce cyanide. *(1 mark)*

Unless a plant has the gene A, it cannot make enzyme A which changes the precurser into linamarin. Even though it has the gene L, without linamarin, cyanide will not be made.

> The candidate has made it clear that without allele A linamarin is not made and would be awarded the mark. She has, however, used the term 'gene' rather than 'allele'. Be sure of the difference between the two terms. A gene is a section of DNA at a particular locus (position) on a chromosome, whilst an allele is a variety or type of a gene.

b) A clover plant has the genotype **AaLl**.

(i) Give the genotypes of the male gametes which this plant can produce. *(1 mark)*

AL, Al, aL, al

> Again this is the correct answer, but be sure that you indicate the different gametes clearly. It is worth getting into the habit of putting each into a circle. Here the candidate has separated each gamete by a comma, but if a series of letters are close together on a line, the examiner may not be sure that you know the difference between them.

(ii) Explain how meiosis results in this plant producing gametes with these genotypes. *(2 marks)*

During meiosis chromosomes come together and swap sections. This is called crossing over and leads to variation. Independent segregation also takes place which means that any gene can come together in the gamete with any other gene.

> This is not a clear answer. The homologous chromosomes do pair during the prophase of the first division of meiosis and their consequential separation leads to only one member of the pair of alleles being present in a gamete. This candidate has not made it clear what pairs with what. This process is called 'segregation', and leads to either allele a or allele A being found in one gamete; and the same principle applies for allele L and allele l. Segregation of one pair of homologous chromosomes takes place independently of any other pair. This process is called 'independent assortment' and leads to either allele a/A going with allele l/L in a gamete. 'Independent segregation' confuses the two processes.
>
> 'Crossing over' has nothing to do with this, as the genes are found on two different chromosomes and, if mentioned, as in this answer, it would lead to the cancellation of the 'independent assortment' marking point.

c) Two plants, heterozygous for both of these pairs of alleles, were crossed. What proportion of the plants produced from this cross would you expect to be acyanogenic but able to produce linamarin? Use a genetic diagram to explain your answer. *(3 marks)*

AaLl × AaLl

	AL	*Al*	*aL*	*al*
AL	AALL	AALl	AaLL	AaLl
Al	AALl	AAll	AaLl	Aall
aL	AaLL	AaLl	aaLL	aaLl
al	AaLl	Aall	aaLl	aall

3 out of 16

This answer would gain two of the three marks. The candidate has correctly derived the gametes of both parents and has carefully filled in the contingency table to show the genotypes of the offspring. The proportion of plants which are acyanogenic but able to produce linamarin is also correct, but she has not identified the genotypes of those plants in the table. AAll, from the third column, third row; Aall from the fifth column, third row and Aall from the third column, fifth row need to be highlighted in some way to prove that she has done more than guessed at an answer.

In an investigation, cyanogenic and acyanogenic plants were grown together in pots. Slugs were placed in each pot and records were kept of the number of leaves damaged by the feeding of the slugs over a period of 7 days. The results are shown in the table.

	Undamaged	Damaged
Cyanogenic	160	120
Acyanogenic	88	192

d) A χ^2 was carried out on the results.

(i) Suggest the null hypothesis that was tested. *(1 mark)*

There was no difference between the damaged and undamaged plants

The null hypothesis does state that 'there will be no difference….' but not between damaged and undamaged plants. There is a difference. Some are damaged and some are not! The null hypothesis should state that there is no difference in damage between the two varieties of plants: cyanogenic and acyanogenic.

(ii) χ^2 was calculated. When this value was looked up in a table, it was found to correspond to a probability of less than 0.05. What conclusion can you draw from this? *(3 marks)*

If the probability is less than 0.05 then the difference is not just due to chance. The null hypothesis is not true and there must be a difference between the damaged and undamaged plants.

This question is worth 3 marks but this candidate has not really stretched herself to look for those marks. The initial error from part d) (i) about what null hypothesis is being tested is evident again, but examiners try never to carry errors forward into subsequent sections of a question. 0.05 is the probability normally accepted by biologists as being the critical value. So she is correct to write that, at a lower level of probability, the difference is not just due to chance. However, the null hypothesis cannot be untrue unless the probability is zero. The data only allow us to reject (or for a probability greater than 0.05, to accept) the null hypothesis. The third mark was given for relating the data to the situation i.e. being cyanogenic does appear to protect those plants from being damaged by slugs.

A second investigation was carried out in a field of grass which had been undisturbed for many years. The table shows the population density of slugs and the numbers of cyanogenic and acyanogenic clover plants at various places in the field.

Population density of slugs	Number of acyanogenic clover plants per m^2	Number of cyanogenic clover plants per m^2
Very low	26	10
Low	17	26
High	0	10
Very high	0	5

e) Explain the proportions of the two types of clover plant in different parts of the field. *(4 marks)*

When there is a high or very high population of slugs there are no acyanogenic clover plants. These plants must have been eaten by slugs. When the number of slugs is low some acyanogenic plants survive and when the number of slugs is very low lots of them survive.

Cyanogenic plants live in all the areas regardless of the number of slugs. The numbers in the sample are very low and maybe the experiment was not repeated. If the investigation was repeated a number of times perhaps the scientist would have found more plants and would have a different pattern. Perhaps these are anomalous results due to a poorly planned experiment.

The data in the table show the population density of slugs and the question asks you to explain why the two different types of plant are distributed unevenly. With the information in the sections of the question that precedes this, it ought to have been clear that cyanide released by cyanogenic plants limits slug damage. Therefore, when there is a high population of slugs, they will eat the other variety of plant, which has no such protection. When there is a low population of slugs, both variety survive, as there is no selection pressure on either form. To criticise the data or the method used to collect it, as in this answer, may be valid, but is not answering the question set.

(AQA 2003)

EXAMINATION QUESTIONS

1 The drawing shows the chromosomes from a cell during meiosis.

a) Name the phase of meiosis shown in the drawing.
Give evidence for your answer. *(2 marks)*

b) What is the haploid chromosome number in this species? *(1 mark)*

c) At the time shown in the diagram, this cell contained 8 picograms of DNA. How much DNA would be present in each gamete produced from this cell? *(1 mark)*

d) In gamete production, what is the advantage of changing diploid cells into haploid cells? *(1 mark)*

(AQA 2002)

2 The inheritance of body colour in fruit flies was investigated. Two fruit flies with grey bodies were crossed. Of the offspring, 152 had grey bodies and 48 had black bodies.

a) Using suitable symbols, give the genotypes of the parents. Explain your answer.
Genotypes:
Explanation: *(2 marks)*

b) (i) Explain why a statistical test should be applied to the data obtained in this investigation. *(2 marks)*

(ii) The chi-squared (χ^2) test was applied to the data obtained. The formula is given below.

$$\chi^2 = \sum \frac{(O - E)^2}{E}$$

Use your formula to determine the value of χ^2 for the results of this investigation. Show your working. *(3 marks)*

(ii) The null hypothesis in this investigation predicted that there would be no difference between the observed and expected values. Use the table to determine whether this hypothesis can be supported.
Explain how you arrived at your answer.

Degrees of freedom	Probability value					
	0.99	**0.95**	**0.1**	**0.05**	**0.01**	**0.001**
1	0.0002	0.0039	2.71	3.84	6.63	10.83
2	0.020	0.103	4.61	5.99	9.21	13.82
3	0.115	0.352	6.25	7.81	11.34	16.27
4	0.297	3.711	7.78	9.49	13.28	18.47

c) A species of insect, only found on a remote island, has a characteristic controlled by a pair of codominant alleles, C^M and C^N.

(i) What is meant by *codominant?* *(1 mark)*

(ii) There were 500 insects in the total population. In this population, 300 insects had the genotype $C^M C^M$, 150 had the genotype $C^M C^N$ and 50 had the genotype $C^N C^N$. Calculate the *actual* frequency of the allele C^N by using these figures. Show your working. *(2 marks)*

(iii) Use your answer to c) (ii) and the Hardy-Weinberg equation to calculate the number of insects that would be *expected* to have the genotype $C^N C^N$. *(3 marks)*

(AQA 2003)

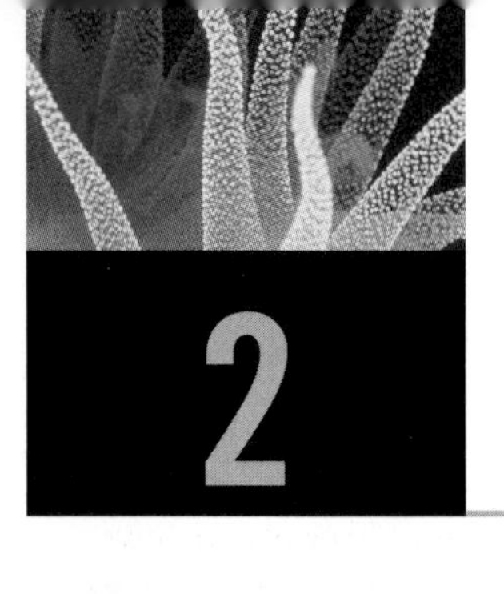

2 Variation and selection

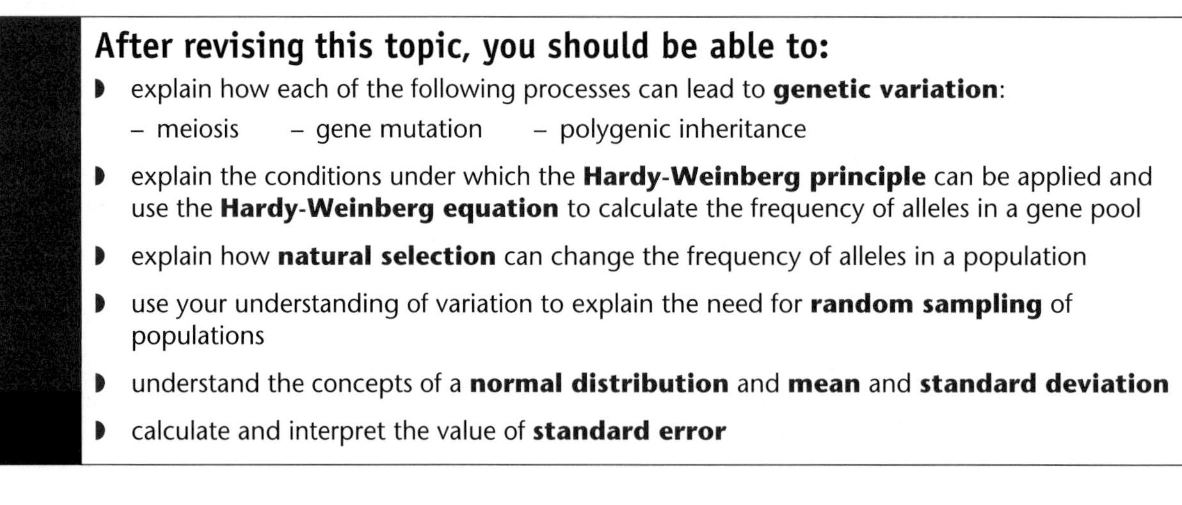

After revising this topic, you should be able to:

- explain how each of the following processes can lead to **genetic variation**:
  – meiosis – gene mutation – polygenic inheritance
- explain the conditions under which the **Hardy-Weinberg principle** can be applied and use the **Hardy-Weinberg equation** to calculate the frequency of alleles in a gene pool
- explain how **natural selection** can change the frequency of alleles in a population
- use your understanding of variation to explain the need for **random sampling** of populations
- understand the concepts of a **normal distribution** and **mean** and **standard deviation**
- calculate and interpret the value of **standard error**

The causes of variation

There is **variation** between the individuals of any natural population. This variation is caused by an **interaction between the environment and the genotype** (Figure 2.1).

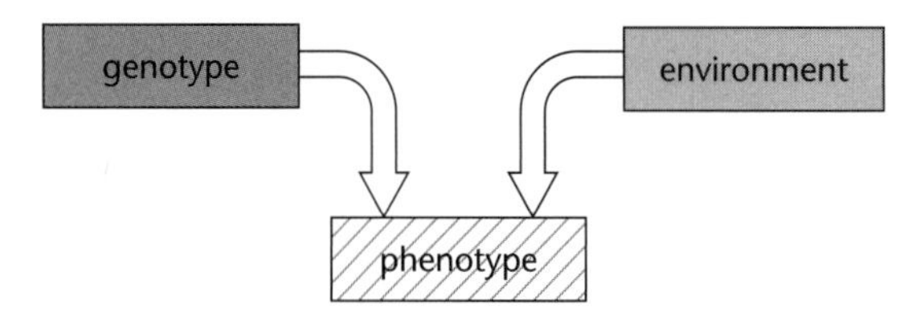

Figure 2.1 Phenotype results from an interaction of genotype and environment.

1 How does the genotype exert its effect on the phenotype?

Genetic variation between individuals can arise through the effects of **meiosis, gene mutation** and **polygenic inheritance**. Table 2.1 summarises how these processes can generate genetic variation.

EXAMINER'S TIP

Examination questions often test your understanding of specific terminology. Make sure you are clear about the difference between:

- genotype and phenotype
- gene and allele

Candidates often have trouble with the term 'allele'. A simple definition that is adequate for one mark is: 'an allele is one of two or more alternative forms of a gene'.

Source of variation	Explanation
Random assortment of chromosomes during meiosis 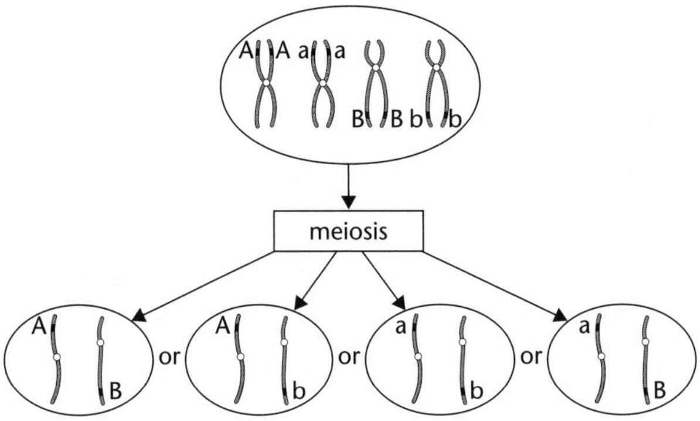	◆ Homologous chromosomes can each carry a **different allele** of the same gene. When these homologous chromosomes are separated during meiosis, half of the daughter cells will carry one allele and half will carry the other allele: they are genetically different. ◆ If each of the members of the 23 pairs of chromosomes in a human cell carried different alleles at just one gene position, random assortment of these chromosomes during meiosis could produce 2^{23} genetically different daughter cells (gametes). ◆ Since **fertilisation of human gametes is random**, the potential variation in the offspring of one pair of humans is enormous.
Crossing over between members of a single pair of homologous chromosomes during meiosis 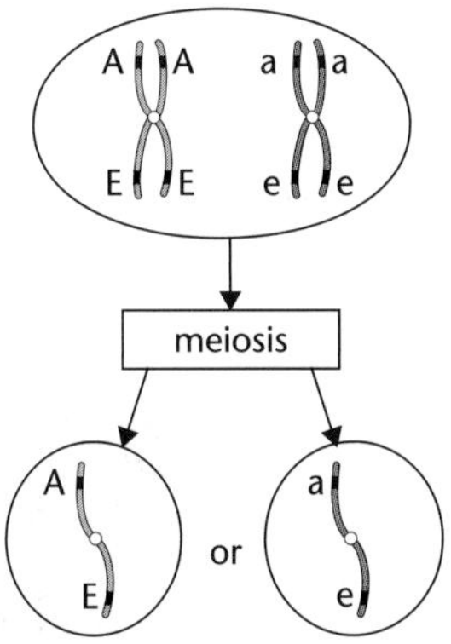 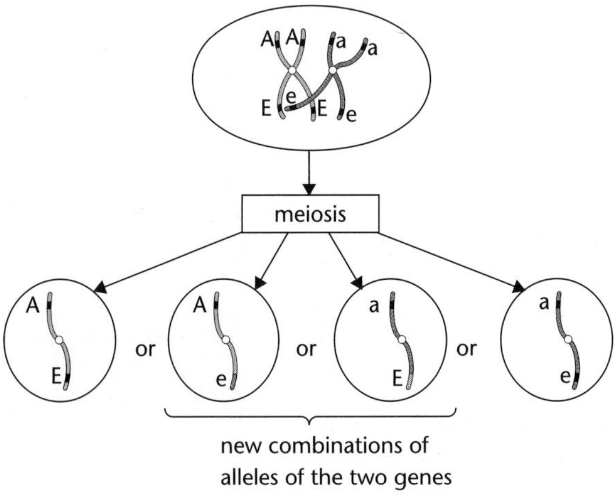	**Chiasmata** often occur during meiosis (see chapter 1). These result in the formation of new genetic combinations on individual chromosomes.

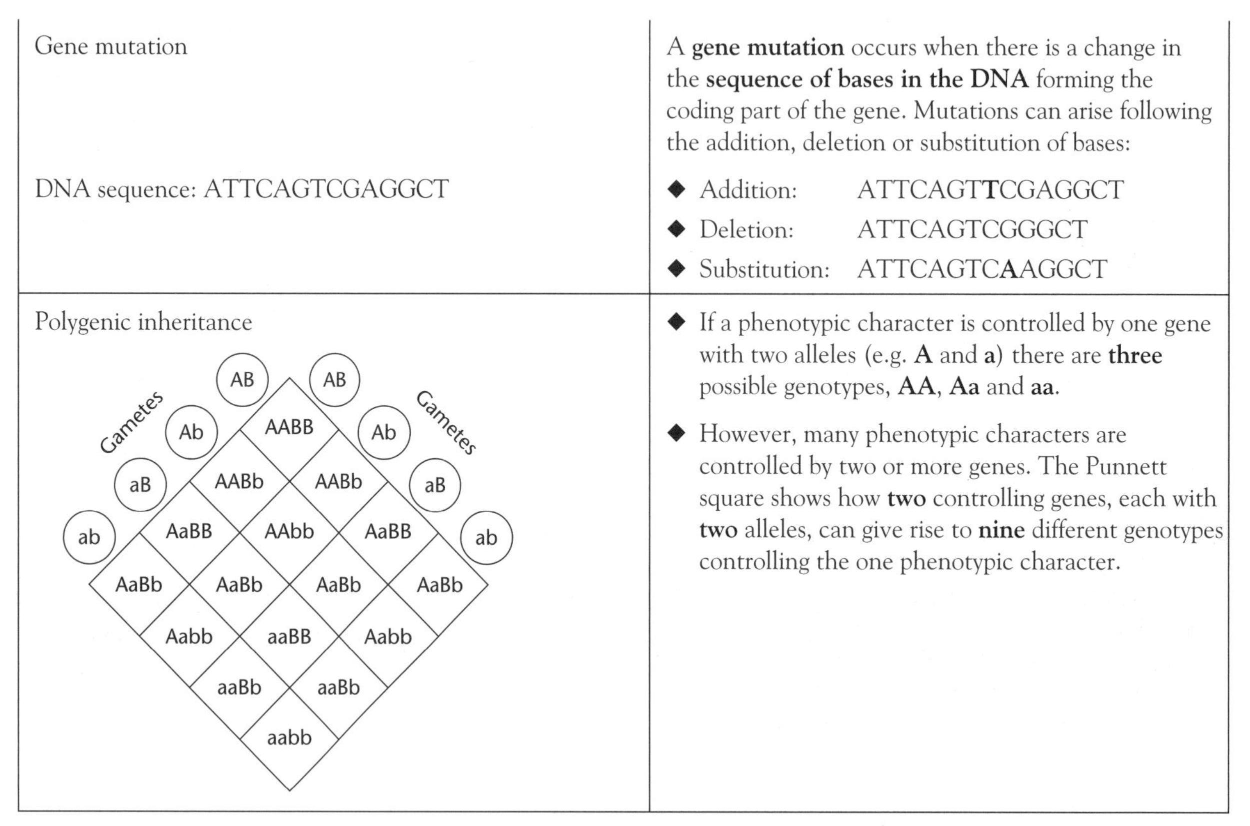

Gene mutation DNA sequence: ATTCAGTCGAGGCT	A **gene mutation** occurs when there is a change in the **sequence of bases in the DNA** forming the coding part of the gene. Mutations can arise following the addition, deletion or substitution of bases: ◆ Addition: ATTCAGT**T**CGAGGCT ◆ Deletion: ATTCAGTCGGGCT ◆ Substitution: ATTCAGTC**A**AGGCT
Polygenic inheritance	◆ If a phenotypic character is controlled by one gene with two alleles (e.g. **A** and **a**) there are **three** possible genotypes, **AA**, **Aa** and **aa**. ◆ However, many phenotypic characters are controlled by two or more genes. The Punnett square shows how **two** controlling genes, each with **two** alleles, can give rise to **nine** different genotypes controlling the one phenotypic character.

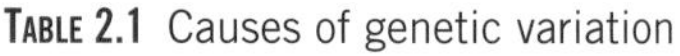
TABLE 2.1 Causes of genetic variation

2 Give **two** ways in which meiosis contributes to genetic variation.

Allele frequencies in populations: the Hardy-Weinberg principle

The **Hardy-Weinberg principle** involves two concepts:

- the **gene pool** – the **total number** of **alleles of a single gene** in the genotypes of individuals in a population. For example, in a population of 10 000 individuals each with a genotype of **AA**, **Aa** or **aa**, there will be a gene pool of 20 000 alleles.
- **allele frequency** – the **proportion of each allele in the gene pool**. (Frequency is always a fraction of 1 expressed as a decimal, e.g. 15 000 **A** alleles in the above gene pool gives a frequency for the **A** allele of $\frac{15\,000}{20\,000} = 0.75$.)

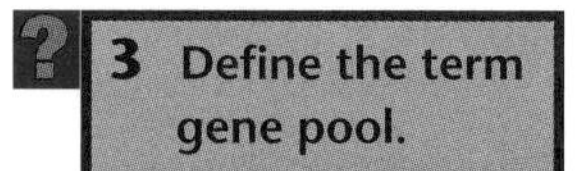
3 Define the term gene pool.

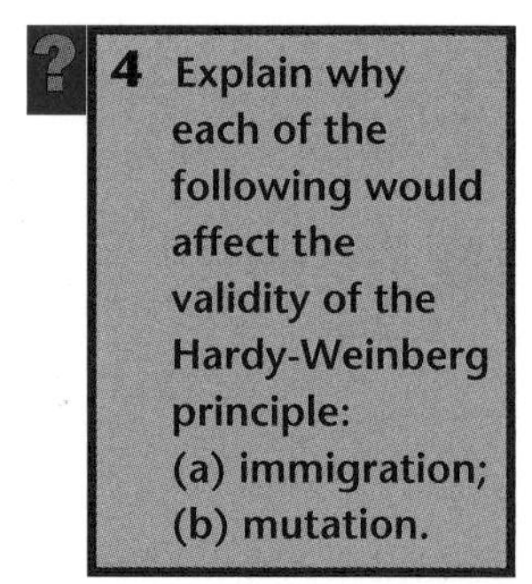
4 Explain why each of the following would affect the validity of the Hardy-Weinberg principle:
(a) immigration;
(b) mutation.

The Hardy-Weinberg principle is based on a calculation about allele frequencies. It states that the **frequencies of the alleles of a particular gene** will **stay constant** from generation to generation provided that:

- the population is **large** (otherwise random changes become important)
- there is **no migration** into the population or out of the population
- there is **random mating** between individuals in the population
- all genotypes are **equally fertile**
- there is no **mutation**

The Hardy-Weinberg equation enables us to **calculate allele frequencies** from the **frequency of genotypes** in a population.

- In an imaginary organism, one gene has two alleles, dominant **A** and recessive **a**.
- In a population of these organisms, three genotypes might occur: **AA**, **Aa** and **aa**.
- The sum total of the **A** and **a** alleles in the genotypes of all the organisms in this population is the gene pool.
- p = the frequency of the **A** allele in this gene pool
  q = the frequency of the **a** allele in this gene pool.
- The frequency of each of the genotypes in this population is:

  AA $= p^2$ **Aa** $= 2pq$ **aa** $= q^2$
- Since p and q are frequencies, $p + q = 1$ and $p^2 + 2pq + q^2 = 1$.

5 If the value of $p = 0.6$, what is the value of q?

6 In a particular population, the phenotype of AA and Aa individuals is identical. The phenotype of aa individuals is different and can be detected. The frequency of aa individuals in a population of 1000 organisms was found to be 0.36.
Use the Hardy-Weinberg equation to calculate
a) the frequency of the Aa individuals
b) the number of Aa individuals in the population.

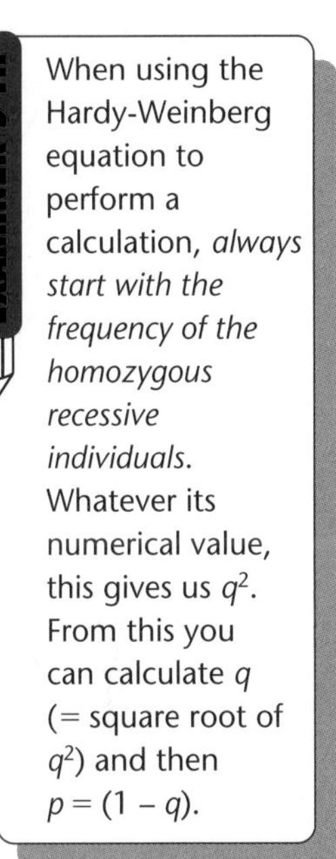

When using the Hardy-Weinberg equation to perform a calculation, *always start with the frequency of the homozygous recessive individuals.* Whatever its numerical value, this gives us q^2. From this you can calculate q (= square root of q^2) and then $p = (1 - q)$.

Changes in allele frequencies in populations: Natural selection

The Hardy-Weinberg principle predicts that allele frequencies will remain unchanged from generation to generation. The principle will **not** hold true if different genotypes have **different fertilities**. Natural selection:

- causes a **change** in the **allele frequency** of a gene in a population
- arises because **individuals have different fertilities**, i.e. individuals with one genotype have more offspring than those of a different genotype. A phenotype which confers the **highest** fertility rate is said to be fit.

Figure 2.2 summarises the process of natural selection.

Natural selection can change allele frequency in three ways. These are shown in Table 2.2. In each case, the upper diagram in each pair shows the frequency distribution of phenotypes before selection and the lower diagram in each pair shows the frequency distribution of phenotypes after selection.

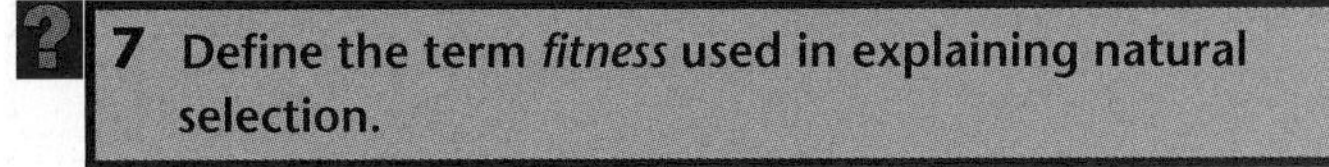

7 Define the term *fitness* used in explaining natural selection.

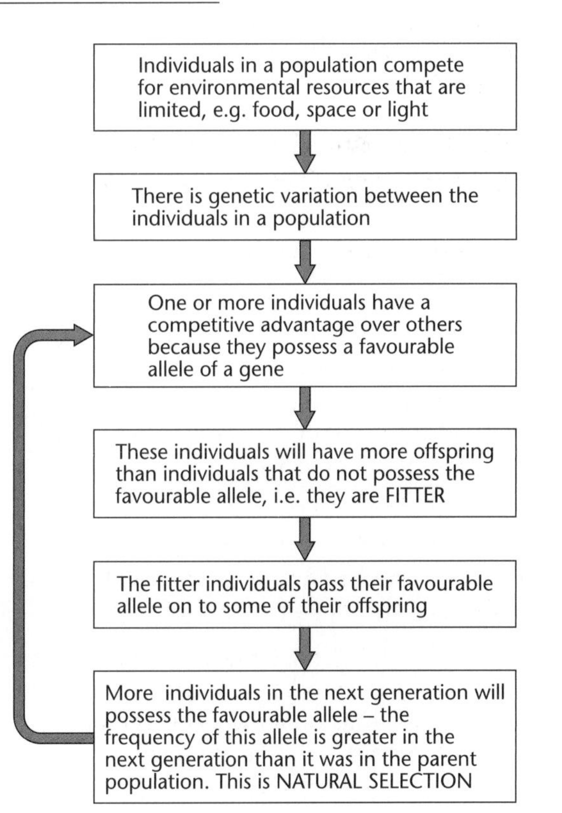

FIGURE 2.2 Natural selection causes the frequency of alleles of a gene to change in a population because of differential fertility resulting from possession of a favourable allele of a gene.

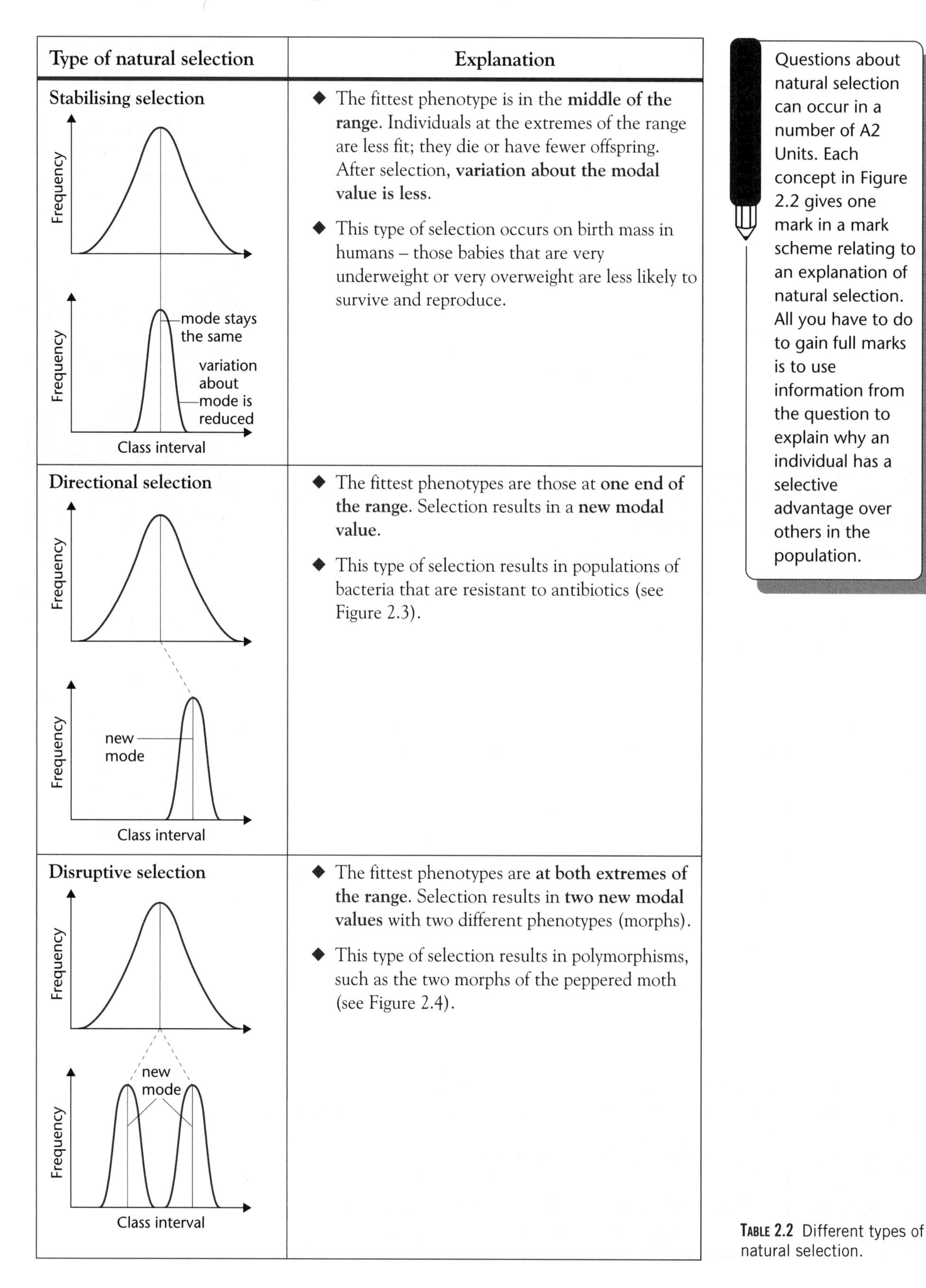

Type of natural selection	Explanation
Stabilising selection	◆ The fittest phenotype is in the **middle of the range**. Individuals at the extremes of the range are less fit; they die or have fewer offspring. After selection, **variation about the modal value is less.** ◆ This type of selection occurs on birth mass in humans – those babies that are very underweight or very overweight are less likely to survive and reproduce.
Directional selection	◆ The fittest phenotypes are those at **one end of the range**. Selection results in a **new modal value.** ◆ This type of selection results in populations of bacteria that are resistant to antibiotics (see Figure 2.3).
Disruptive selection	◆ The fittest phenotypes are **at both extremes of the range**. Selection results in **two new modal values** with two different phenotypes (morphs). ◆ This type of selection results in polymorphisms, such as the two morphs of the peppered moth (see Figure 2.4).

Questions about natural selection can occur in a number of A2 Units. Each concept in Figure 2.2 gives one mark in a mark scheme relating to an explanation of natural selection. All you have to do to gain full marks is to use information from the question to explain why an individual has a selective advantage over others in the population.

TABLE 2.2 Different types of natural selection.

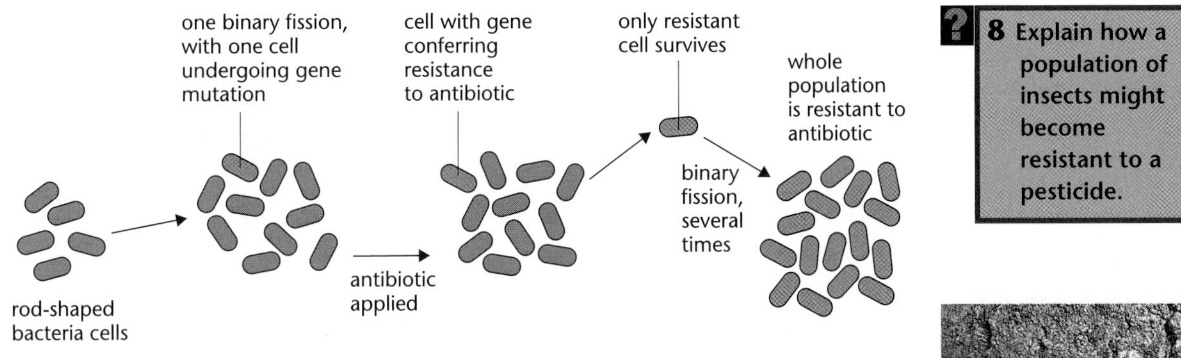

Figure 2.3 Directional selection in a population of bacteria. A chance mutation enabled one bacterial cell to survive the effects of an antibiotic. As all of its offspring are clones, they inherited the new allele for antibiotic resistance. The frequency of the new allele changed dramatically from almost zero to 1.0 in a single generation.

8 Explain how a population of insects might become resistant to a pesticide.

Figure 2.4 Disruptive selection has produced two morphs of the peppered moth (*Biston betularia*). The mottled morph is at an advantage in unpolluted areas where it is less easily detected by predatory birds. The black morph is at a similar advantage in polluted areas.

NATURAL SELECTION IN HUMANS

Sickle-cell anaemia is common in many parts of the world. It results from a mutation in a gene that codes for one of the polypeptides in haemoglobin. The alleles involved are:

- Hb^A – codes for haemoglobin-A
- Hb^S – codes for a form of haemoglobin-A that reduces the oxygen-carrying capacity of haemoglobin

Table 2.3 shows the genotypes involving the haemoglobin-A gene. Notice the genotypes have two effects on the phenotype: one on the oxygen-carrying capacity of the haemoglobin and one on the person's susceptibility to malaria.

Genotype	Oxygen-carrying capacity	Susceptibility to malaria
Hb^A Hb^A	All red blood cells carry oxygen well	Susceptible to malaria
Hb^A Hb^S	Half red blood cells carry oxygen well and half carry oxygen less well. This person suffers mild anaemia.	Less susceptible to malaria
Hb^S Hb^S	All red blood cells carry oxygen less well. A person with this genotype suffers severe anaemia	Less susceptible to malaria

Table 2.3 The different genotypes resulting from the alleles controlling the production of human haemoglobin-A

9 Prior to medical advances, natural selection operated on the sickle-cell trait in human populations. Explain why natural selection would have resulted in:

a) a low frequency of the Hb^S allele in populations living in regions where malaria was rare;

b) a high frequency of individuals with the Hb^A Hb^S phenotype in populations living in regions where malaria was endemic.

Frequency distributions

We collect **data** when we carry out investigations. Table 2.4 shows how we might record our raw data if we were observing the number of times a female or male bird entered a nest in a given time interval. Table 2.5 shows how we might collect our raw data if we were recording the heights of students in a class.

Female entered nest	*Male entered nest*
卌 |||	卌 卌 卌 ||
Female entered 8 times	*Male entered 17 times*

Table 2.4 Raw data from an investigation into the behaviour of female and male birds

- The data in Table 2.4 are **discrete**, i.e. there are **distinct categories**: in this case two – male or female.
- The data in Table 2.5 are **continuous**, i.e. there are no **distinct categories**.

These differences become obvious in Figure 2.5, when we plot discrete and continuous data in frequency distributions.

Class interval/mm	*Tally count of number of students*	*Number of students*
1480–1499			1
1500–1519	|||	3
1520–1539	||||	4
1540–1559	卌		6
1560–1579	卌 ||	7
1580–1599	卌 卌	10
1600–1619	卌 |||	8
1620–1639	||||	4
1640–1659	|||	3
1660–1679	||	2
1680–1699			1

Table 2.5 Raw data from an investigation into the height of students in a class

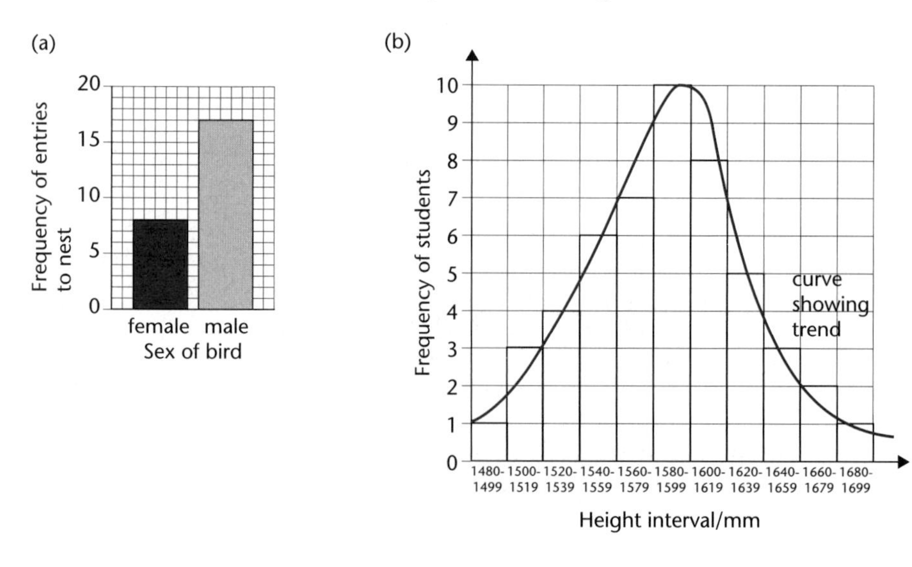

Figure 2.5 Frequency distributions of the data in (a) Table 2.4 and (b) Table 2.5. The data from Table 2.4 are discrete. The coloured line in (b) shows the general trend of the continuous data in Table 2.5.

NORMAL DISTRIBUTION

A **normal distribution** is a special type of frequency distribution. Figure 2.6 shows a graph of a normal distribution – a NORMAL DISTRIBUTION CURVE. This has important mathematical properties:

- its most frequent value (**mode**), middle value (**median**) and arithmetic average value (**mean**) are the **same**;
- it is **symmetrical** – 50% of its values are above the mean and 50% of its values are below the mean.

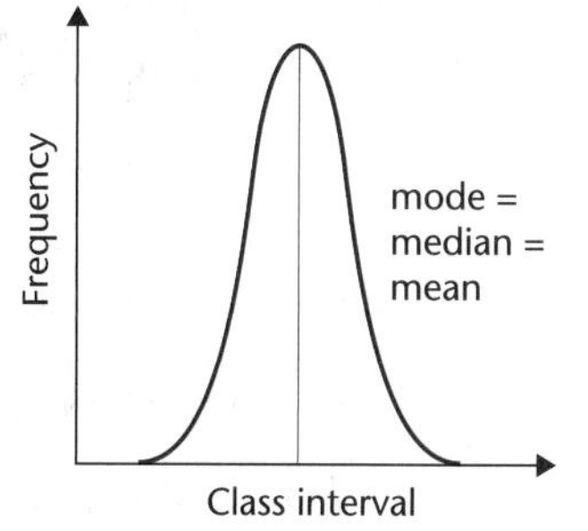

FIGURE 2.6 A normal distribution curve. This is a special type of frequency distribution curve in which the mean, median and mode have the same value.

Figure 2.7 shows two normal distributions. They have the same mean, but the spread of data about the mean is different. We measure the spread of data about a mean value using the **standard deviation**. Put simply, graph (a) in Figure 2.7 has a greater standard deviation than has graph (b).

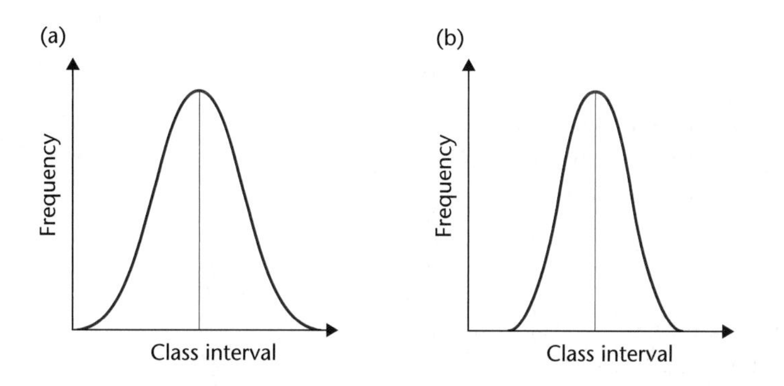

FIGURE 2.7 Two normal distribution curves. Curve (a) has a greater standard deviation than curve (b).

Sampling populations

When we sample populations, we measure data about only a small part of the entire population. Thus, we only **estimate** the true mean and standard deviation of the whole population. Our **measured estimate of the standard deviation** is called the **standard error** (SE). This gives us another important feature of a normal distribution curve, shown in Figure 2.8.

- 95% of the values in a normal distribution curve lie in the range of mean ± 1.96 SE.

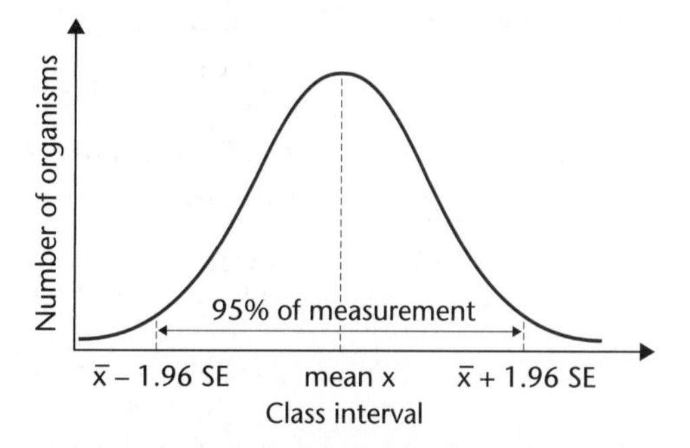

FIGURE 2.8 A distribution curve obtained from a sample from a whole population. The mean of the sample, $\bar{x}$, is an estimate of the mean of the whole population. 95% of the measurements lie within the range of mean ± 1.96 × standard error.

The standard error is an important concept. Suppose we have measured two sets of data and we want to know if they are different. We calculate the mean and standard error for each set of data. If the range (mean ± 1.96 times the SE) of one set of data **overlaps** with the calculated range for the second set of data, they are **not significantly different** from each other. If, however, the two ranges (mean ± 1.96 times the SE) do **not overlap**, the two means are SIGNIFICANTLY DIFFERENT from each other.

- SIGNIFICANTLY DIFFERENT means that there is **less than a one-in-twenty** chance of their differences being due only to chance.
- A one-in-twenty chance can be expressed as a 5% chance or $p = 0.05$.

Using standard error is a simple way to demonstrate Skill E in your A2 coursework. You calculate standard error by using the formula:

$$\text{STANDARD ERROR (SE)} = \frac{s}{\sqrt{n}}$$

In this formula, n is the number of measurements made and s is the standard deviation. You calculate the value of **s** using the formula:

$$\text{Standard deviation, } s = \sqrt{\frac{\Sigma(x - \bar{x})^2}{n - 1}}$$

In this formula:

- x = each individual measurement
- $\bar{x}$ is the mean of all the values of x
- Σ means the sum of all the values in the brackets
- n = number of measurements made

Since our ability to evaluate data from investigations depends on our interpretation of standard error, it is vital that **samples reflect the true population** as closely as possible. In your coursework, you will design your investigation to do this by:

- collecting **more than one** sample. This will involve collecting several samples during fieldwork, or by repeating an experiment several times in a laboratory-based investigation.
- ensuring there is **no personal bias** in the samples you take. In fieldwork, this might involve producing a grid over the area to be sampled and choosing the coordinates of squares to be sampled using a table of random numbers.

? **10** The mean heights of two samples of plants were calculated. The ranges of mean height ± 1.96 × SE of these populations did not overlap. What does this tell us about the differences in the mean heights of the two samples of plants?

EXAMINER'S TIP

You will not be asked to calculate either standard deviation or standard error in examination papers. You will only perform these calculations in your A2 coursework.

WORKED EXAM QUESTIONS

1 Huntington's disease is a human inherited condition resulting in gradual degeneration of nerve cells in the brain. It is caused by a dominant allele, but usually no symptoms are evident until the person is at least 30 years old. It is very rare in most populations. However, in one isolated area in Venezuela, 48% of the population possess a genotype which gives rise to Huntington's disease. Many of the inhabitants of this area can trace their origins to a common ancestor 200 years ago.

a) Use the Hardy-Weinberg equation to estimate the percentage of this Venezuelan population which is heterozygous for Huntington's disease. Show your working. *(3 marks)*

48% as a frequency is 0.48
frequency of homozygous recessives = 1 – 0.48 = 0.52
frequency of recessive allele is square root of 0.52 = 0.72
∴ frequency of dominant allele = 1 – 0.72 = 0.28
∴ frequency of homozygous individuals = 2pq = 2 x 0.28 × 0.72 = 0.403
Answer 40 %

This candidate has shown his working very clearly and seems to have no problem using frequencies and percentages. Correctly, he started with the homozygous recessive individuals and used information in the question to calculate their frequency. Knowing this, he correctly used $p + q = 1$ and $p^2 + 2pq + q^2 = 1$ to calculate the frequency of the heterozygotes ($2pq$). The mark scheme was: correct use of $p + q = 1/p^2 + 2pq + q^2 = 1$; identifying that frequency of heterozygotes was $2pq$/use of appropriate numbers; answer correctly calculated.

b) Suggest why: *(3 marks)*

(i) there is such a high incidence of Huntington's disease in this population;

the gene for Huntington's disease is present in larger than normal amounts in that area's gene pool & is more likely to be present in a person's DNA.

This answer scores no marks. The candidate has not given a reason to explain why there is a high incidence of the disease. Marks were awarded for: common ancestor; in-breeding/no immigration/no migration/genetic isolation/small gene pool; high probability of mating with a person having the dominant allele.

(ii) Huntington's disease has not been eliminated from this population by natural selection.

As it is not evident until the person is 30 years old & may already have had children by the time the symptoms begin to appear, the infected person is at no disadvantage to compete and is therefore not eliminated.

This answer is not particularly well expressed but gains two marks. Marks were available for: reproduction occurs before symptoms of disease are apparent; possessors of the dominant allele are not at a survival/selective disadvantage; a genetic argument, e.g. Hh × hh → 50% of offspring affected/Hh × Hh → 75% of offspring affected.

(AQA 2002)

EXAMINATION QUESTIONS

1 One variety of the plant false flax grows in fields. A second variety grows on roadsides. False flax plants from the two habitats differ in the size of their seeds. In an investigation, seeds were collected from the two habitats and their diameters were measured. The results are shown in the graph.

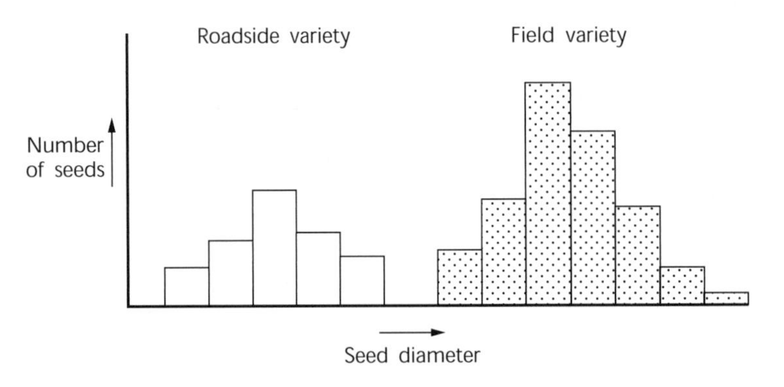

a) The two varieties of false flax have evolved by disruptive selection.

(i) Use information in the graph to explain what is meant by *discontinuous variation*. *(1 mark)*

(ii) Suggest how disruptive selection might have given rise to the distribution of seed diameter as shown in the graph. *(2 marks)*

b) Describe how you could show that both varieties of false flax belong to the same species. *(2 marks)*

(AQA 2003)

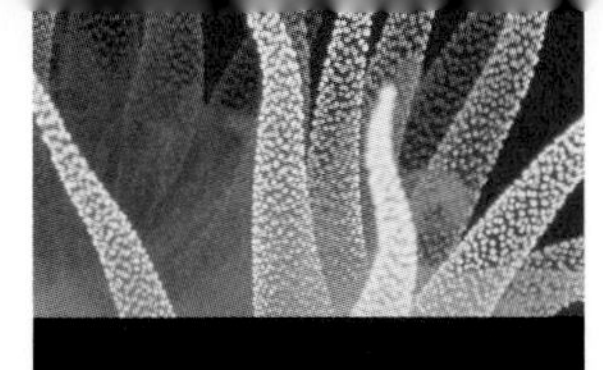

3 The evolution and classification of species

After revising this topic, you should be able to:

- define, and show an understanding of, the term *species*
- explain the importance of **reproductive isolation** in the formation of new species
- distinguish between **allopatric** speciation and **sympatric** speciation
- correctly **group** species into the following larger categories: genus, family, order, class, phylum and kingdom
- name, and give distinguishing features of, the **five kingdoms** of organisms

The concept of species

Two organisms belong to the **same species** if:

- they have the **potential to breed** with each other, and
- their **offspring are fertile**.

Sometimes, this definition is not helpful. For example, we cannot test the two criteria if the organisms are only known from fossil records, or if they do not reproduce sexually. In these cases, we have to look for **lack of variation** in features such as structure, biochemistry, development or ecological niche.

EXAMINER'S TIP

A question asking for a definition of the term *species* is likely to carry two marks. These will be awarded for: members of the same species have the potential to breed; and (breeding) produces fertile offspring. Make sure you can recall this definition for an easy two marks.

Speciation

The formation of **new** species from **existing** species is an important concept in biology.

HYBRIDISATION AND POLYPLOIDY

In plants, this is known to have arisen through **hybridisation and polyploidy**. Table 3.1 explains how these processes led to the formation of new species of wheat.

Process	Explanation	Diagrammatic representation
Hybridisation	◆ The hybrid was formed by a **cross** between: – an ancestral species of wheat, called *Triticum urartu*, with a diploid genotype (2n) represented as **TT** – a species of grass, called *Aegilops speltoides*, with a diploid genotype (2n) represented as **AA** ◆ The new hybrid inherited a haploid set of chromosomes from each parent, so had a diploid genotype (2n) of **AT**	*Triticum urartu* (TT) × *Aegilops speltoides* (AA) ↓ sterile hybrid (AT)

Polyploidy	A **random mutation** caused one sterile hybrid to **double** its chromosome number (4n). This made the polyploid plant **fertile**. This plant is commonly known as durum wheat and is used to make pasta.	Sterile hybrid (AT) | *Triticum turgidum* (AATT)

TABLE 3.1 The formation of a new species of wheat through hybridisation and polyploidy. The symbols **A** and **T** represent the haploid chromosome complement of *Aegilops speltoides* and *Triticum urartu*, respectively.

1 Suggest why the hybrid in Table 3.1 was sterile whereas the tetraploid AATT was fertile.

2 Use the information in Figure 3.1 to:
a) name the process by which cultivated cotton was produced from wild cotton;
b) give the chromosome number in vegetative cells of wild (uncultivated) cotton plants.

FIGURE 3.1 These cultivated varieties of cotton (52 chromosomes) and tobacco (24 chromosomes) are tetraploid (4n).

REPRODUCTIVE ISOLATION

Outside the plant kingdom, polyploidy is rare. Table 3.2 shows the more usual method of speciation.

Process	Explanation
Isolation	A population of organisms is **separated** into two, or more, **groups**.
Reproductive barriers	**Breeding** between the isolated groups is **prevented**.
Genetic changes	Through **mutation** and **natural selection**, the frequency of alleles changes differently in the isolated groups.
New species formation	If the genetic changes are sufficiently great, the members of the isolated groups become so **different** that **interbreeding** to produce fertile offspring becomes **impossible**.

TABLE 3.2 Speciation occurs following genetic changes in populations that have become reproductively isolated from each other.

Reproductive isolation is critical to speciation and can result in two ways. Table 3.3 summarises them.

Type of speciation	Cause of reproductive isolation	Explanation
Allopatric speciation	**Geographical barriers**	◆ **Geographical** isolation prevents organisms from the different groups coming into contact. ◆ Barriers include oceans and canyons.
Sympatric speciation	**Time** when organisms are active, different ecological **niches** or **physiological** barriers	Organisms which are **active at different times**, **breed at different times**, occupy **different niches** in the same habitat or are **physiologically incompatible**, cannot interbreed to form fertile offspring.

TABLE 3.3 The reproductive isolation of two, or more, groups of a single species can arise from geographical or other forms of separation.

4 A desert separates two populations of mice and two populations of predatory birds. Explain why the desert might lead to the formation of new species of mice but not of predatory birds.

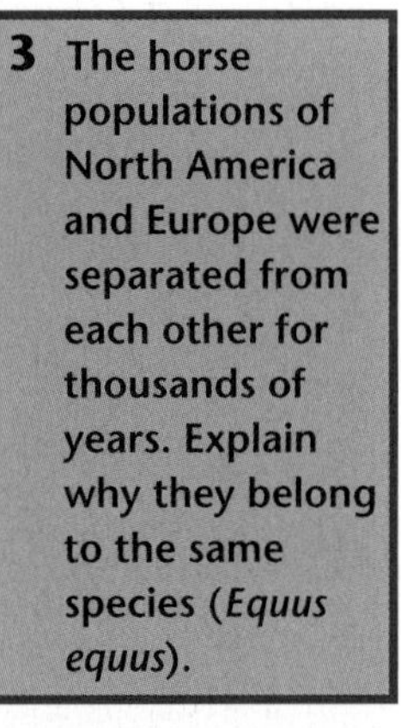
3 The horse populations of North America and Europe were separated from each other for thousands of years. Explain why they belong to the same species (*Equus equus*).

Classification of species into larger groups

Biologists group **similar** species together in a **larger group** (category), called a **genus**. In turn, genera (plural of *genus*) are grouped together into even larger categories and so on. Figure 3.2 shows the most frequently used biological categories.

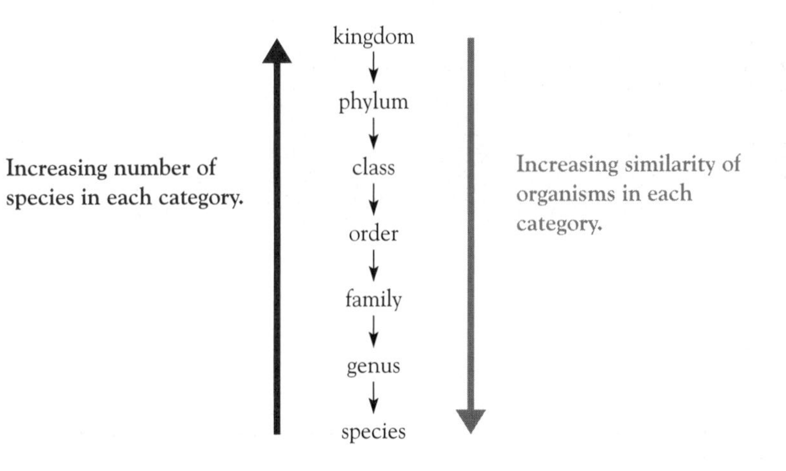

FIGURE 3.2 Species are grouped into a category of similar organisms, called a genus. In turn, each category is placed in a larger category. The kingdom is the largest category of all.

Sometimes, a mnemonic is helpful in remembering a sequence. Candidates often forget the sequence of categories in biological classification. A mnemonic can help you to remember the correct sequence. For example, **King Philip Came Over For German Sausages** might help you to remember the sequence of **kingdom, phylum, class, order, family, genus and species**.

The **biological name** of each species is derived from its **genus** and **species names** (a **binomial**).

- Name of genus takes an upper case letter.
- Name of species takes a lower case letter.
- The full name is printed in italics or underlined if handwritten e.g., the binomial for humans is *Homo sapiens* or, if handwritten, Homo sapiens.

5 In which genus do humans belong?

The five-kingdom classification of organisms

All living organisms are classified into five kingdoms: Prokaryotae, Protoctista, Fungi, Plantae and Animalia. Table 3.4 summarises the distinguishing characteristics of each of these kingdoms.

Name of Kingdom	Examples	Distinguishing characteristics
Prokaryotae	Bacteria	◆ Cells are prokaryotic (see AS notes) ◆ Autotrophic (chemosynthesis or photosynthesis) and heterotrophic ◆ Cells have cell wall made of peptidoglycans/murein
Protoctista	*Plasmodium vivax* (the cause of malaria), seaweeds	◆ Eukaryotic (see AS notes) ◆ Autotrophic (e.g. seaweeds) and heterotrophic (e.g. *Plasmodium*) ◆ Cells of some have a cellulose cell wall (e.g. seaweeds) whilst cells of others do not (e.g. *Plasmodium*) ◆ Organisms are classed here if they do not fit any other kingdom
Fungi	Moulds, yeast, mushrooms	◆ Eukaryotic ◆ Usually heterotrophic ◆ Cells have walls made of chitin ◆ Usually body is a mass (mycelium) of thread-like filaments (hyphae) ◆ Reproduce by forming resistant spores

Plantae	Mosses, ferns, flowering plants	◆ Eukaryotic ◆ Multicellular ◆ Autotrophic (photosynthesis) ◆ Cells have wall made of cellulose ◆ Have a complex life cycle with a sexually reproducing adult stage and an asexually reproducing adult stage
Animalia	Sea anemones, earthworms, insects, snails, fish, humans	◆ Eukaryotic ◆ Multicellular ◆ Heterotrophic ◆ Cells lack walls ◆ Have nervous systems ◆ Embryo has a stage at which it is a hollow ball of cells (the blastula)

TABLE 3.4 The five-kingdom classification of organisms

EXAMINER'S TIP

Although the names of the five kingdoms have Latin endings, you will not be penalised in an examination if you use anglicised versions of the names: prokaryotes, protoctists, fungi, plants and animals.

6 Give two ways in which prokaryotic cells differ from eukaryotic cells.

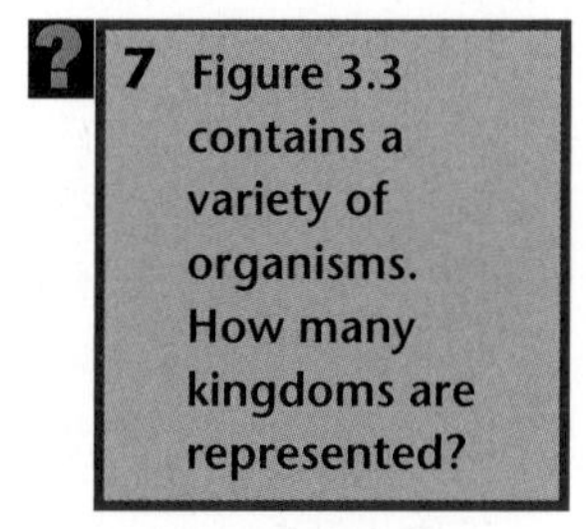

7 Figure 3.3 contains a variety of organisms. How many kingdoms are represented?

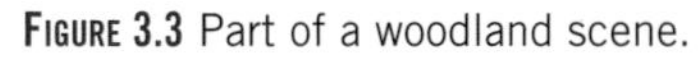

FIGURE 3.3 Part of a woodland scene.

WORKED EXAM QUESTIONS

1 a) The common broomrape, *Orobranche minor*, is an unusual flowering plant. It is a parasite. It has no leaves and does not contain chlorophyll.

(i) Give **one** feature of the common broomrape which is shares with all other plants but does not share with organisms from other kingdoms. *(1 mark)*

It has cellulose cell walls.

> This answer is correct and gains the mark. The candidate could also have written that only plants have large/permanent vacuoles.

(ii) Complete the table to show the classification of the common broomrape. *(2 marks)*

Kingdom	Plant
Phylum	Angiospermaphyta
Class	Dicotyledonae
Family	Scrophulariales
Order	Orobranchaceae
Genus	***Orobranche***
Species	***minor***

> Do not be put off by unfamiliar names in a table like this – you are not expected to know them. The empty boxes for you to fill in are those that you should know. This question is marked by column. If the entire left-hand column is correct, the candidate gains one mark and if the entire right-hand column is correct, the candidate gains a mark.
>
> The candidate has made a mistake in the left-hand column: order should have been above family. The candidate has recalled the name of the kingdom (Plant/Plantae) and has used the information in the first line of the question correctly to identify the name of the genus and species. This answer gains one mark for a correct right-hand column.

b) The common poppy and the long-headed poppy are similar plants. Under natural conditions they sometimes cross to form hybrids between the two. Suggest how you could find out whether the common poppy and the long-headed poppy are different species. *(2 marks)*

I would cross two of the hybrid plants. If they crossed and produced seeds that germinated, they would be the same species. If they did not produce seeds or their seeds failed to germinate, they would be different species of poppy.

> This is an excellent answer. The candidate gained one mark for attempting to cross the two types of poppy and a second mark for realising that they would not produce offspring that would grow.

(AQA 2003)

EXAMINATION QUESTIONS

1 a) What is meant by *reproductive isolation*? *(1 mark)*

b) Explain how geographical isolation can lead to the formation of new species. *(4 marks)*

(AQA 2003)

4 Numbers and diversity

After revising this topic, you should be able to:

- define the terms **ecosystem**, **community**, **population**, **habitat** and **niche**
- show understanding of, and calculate from given data, the **index of diversity**
- relate values of the index of diversity to **biotic** and **abiotic factors**
- explain **ecological succession** from **pioneer species** to **climax community**
- show a critical appreciation of two different ways of investigating numbers – **random sampling using quadrats** and the **mark-release-recapture method**

Ecological terms

Table 4.1 includes the specific terminology you must use in ecology.

Table 4.1 Specific terminology used in ecology

Term	Meaning
Abiotic factors	**Physical and chemical features** of the environment that have an effect on a population of organisms, e.g. temperature, substratum (rock, soil or sand), availability of nitrate ions.
Biotic factors	**Biological features** of the environment that have an effect on a population of organisms, e.g. presence of food, competitors, predators or parasites.
Community	**All** the populations living in a particular place at the same time, e.g., a woodland community includes all the plants and animals living in a wood.
Ecosystem	The community of living organisms **and** the abiotic factors which affect them. A seashore ecosystem includes the physical and chemical aspects of the environment plus the communities living on the seashore.
Habitat	The **place within an ecosystem** in which a population lives.
Population	A group of organisms of the **same species** living in the same place.
Niche	A precise description of the **factors affecting a single population**. This includes the tolerance range of the population to each abiotic factor as well as the nature of biotic factors, such as presence of a suitable food source and presence of predators.

EXAMINER'S TIP

The term *niche* is notoriously difficult to define. Use one of the following definitions.
1. Niche describes the **role** an organism plays in an ecosystem, e.g. a limpet grazes on producers (seaweeds) on a rocky shore and a rabbit grazes on producers (grass) in a grassland.
2. Niche describes the abiotic and biotic factors that **enable a population to survive** in a particular environment.

1 Distinguish between the terms in the following pairs.
a) Abiotic factors and biotic factors. b) Community and ecosystem.

Diversity

The diversity of a community reflects:

- the number of **different species** present
- the number of **individuals** in each species

The **index of diversity** enables us to make an objective assessment of diversity in any community. We calculate the index of diversity using the formula:

$$\text{Index of diversity, } d = \frac{N(N-1)}{\Sigma n(n-1)}$$

where N = total number of individuals of all species present in the community
n = total number of individuals of each individual species

WORKED EXAMPLE

Table 4.2 shows some information about species of butterfly in a meadow. We can calculate the index of diversity of this meadow in the following way.

$$\text{Index of diversity, } d = \frac{N(N-1)}{\Sigma n(n-1)}$$

$$d = \frac{21(21-1)}{3(3-1) + 5(5-1) + 7(7-1) + 2(2-1) + 4(4-1)}$$

$$= \frac{21 \times 20}{(3 \times 2) + (5 \times 4) + (7 \times 6) + (2 \times 1) + (4 \times 3)}$$

$$= \frac{420}{82} = 5.12$$

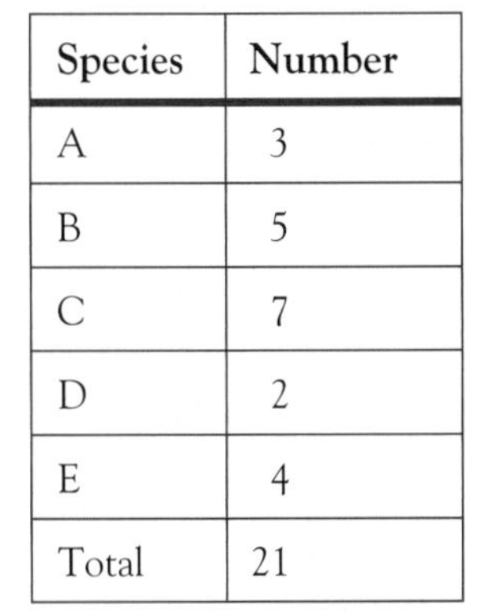

Species	Number
A	3
B	5
C	7
D	2
E	4
Total	21

TABLE 4.2

FIGURE 4.1 A tropical rainforest, such as this one, has a very high index of diversity.

> **?** **2 A tropical rainforest, such as the one shown in Figure 4.1, has a high index of diversity. What does this information tell you about (a) the number of populations and (b) the size of each population within the rainforest?**

Table 4.3 shows how we can make judgements about the harshness and stability of an environment using knowledge of its index of diversity.

Index of diversity	Nature of environment	Explanation
Low value	Unfavourable (e.g., desert, Arctic tundra, upper seashore)	◆ **Few** species present and often populations are **small**. ◆ Generally, **abiotic** factors determine which species are present. ◆ Ecosystems in these environments are usually **unstable**.
High value	Favourable (e.g. tropical rainforest, temperate woodland, lower seashore)	◆ **Many** species present and populations are usually **large**. ◆ Generally, **biotic** factors, such as competition determine which species are present. ◆ Ecosystems in these environments are usually **stable**.

TABLE 4.3 The index of diversity is an indication of the stability of an ecosystem

3 The index of diversity of seaweeds in the sheltered shore in Figure 4.2 is much higher than that in the exposed shore. What does this tell you about the conditions on these two shores?

FIGURE 4.2 (a) A sheltered sea shore and (b) an exposed sea shore.

Ecological succession

Ecosystems **change** over time. These changes are, collectively, called **ecological succession**. Figure 4.3 shows how plant succession might occur on a bare rock surface. Each stage, or **sere**, in the succession results in a **less hostile** environment, as the rock face is weathered and decaying plants provide a nutrient-rich medium in which other plants can grow. As a consequence, diversity **increases**.

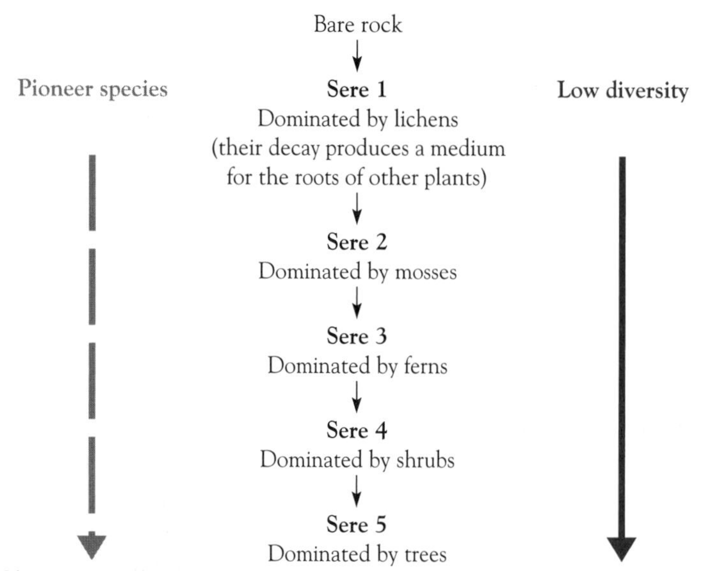

FIGURE 4.3 Succession on a bare rock surface from a pioneer species to a complex ecosystem.

4 Suggest how the plants in each sere shown in Figure 4.3:
a) enable those of the next sere to become established;
b) eliminate those from the previous sere.

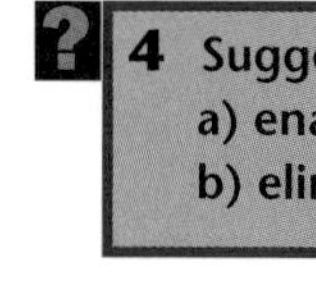

- You are not expected to have specific knowledge of succession in any particular environment. This means that an examiner must give you information in the question and ask you to use this information to test your understanding of succession.
- Always expect examination papers to contain unfamiliar information. It is the only way that examiners can test your ability to use your knowledge and understanding in an unfamiliar context.

Practical methods of investigating numbers and distribution

You probably carried out an ecological investigation in your A2 Biology course. You would have found that you did not have enough time to count and record the positions of all the organisms in the population you were investigating. Instead, you **took samples**.

SAMPLING USING QUADRAT FRAMES

A **quadrat frame** encloses an area of **known dimensions**. Placed over the surface being sampled, we **count the number** of organisms within the frame; or we **estimate the percentage of the area that they cover** within the quadrat (called **percentage cover**). To ensure our samples accurately reflect the true population, we must:

- take a relatively large number of samples
- ensure there is no personal bias in where we place the quadrat. Figure 4.4 shows how we can use measuring tapes and random numbers to ensure our quadrat frames are placed randomly.

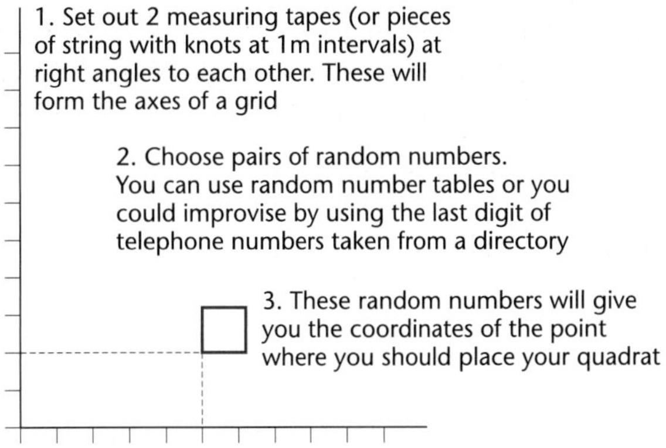

1. Set out 2 measuring tapes (or pieces of string with knots at 1 m intervals) at right angles to each other. These will form the axes of a grid

2. Choose pairs of random numbers. You can use random number tables or you could improvise by using the last digit of telephone numbers taken from a directory

3. These random numbers will give you the coordinates of the point where you should place your quadrat

FIGURE 4.4 Placing quadrats at random. Where possible, a method like this should be used. Throwing the quadrat frame over your shoulder will not result in genuinely random sampling.

SAMPLING USING TRANSECTS

A **transect** is a **line through an area** to be studied. We can take samples at regular intervals along a transect, using a quadrat frame. We generally use a transect where the environment changes gradually, such as the seashore in Figure 4.5.

? **5 Suggest why it is important during mark-release-recapture sampling that we a) mark the animals in a way that is inconspicuous, b) leave a period of time before taking a second sample.**

FIGURE 4.5 This student is investigating the number and distribution of seaweeds on a rocky shore using quadrats along a line transect.

SAMPLING USING THE MARK-RELEASE-RECAPTURE METHOD

Quadrat frames and transects are fine so long as the organisms do not move about whilst we are sampling. The **mark-release-recapture method** allows us to estimate the number of **active animals** in a given area. To perform this method, we:

- capture a number of animals from the population being investigated
- mark these animals in a way that causes no harm and is inconspicuous
- release the marked animals and allow sufficient time for them to disperse back into the population from which they were taken
- take a **second sample** and count the number of animals that are marked and the number that are not marked
- perform the calculation below to estimate the size of the population:

$$\frac{\text{number of marked animals in second sample}}{\text{total number of animals in sample}} = \frac{\text{number of marked animals released}}{\text{total number of animals in population}}$$

$$\text{total number of animals in population} = \frac{\text{number of marked animals released} \times \text{total number of animals in sample}}{\text{number of marked animals in second sample}}$$

WORKED EXAM QUESTIONS

1 The community present in a roadside ecosystem was investigated.

a) Explain what is meant by:

(i) community *(1 mark)*

This means all the different populations of organisms in the ecosystem.

> This answer is correct. One mark would be awarded for all the organisms/ populations in an area but not for all the species.

(ii) ecosystem *(1 mark)*

This means all the abiotic and biotic factors.

> This answer is acceptable and gains one mark. An alternative answer is: 'the habitat plus the community that lives within it'.

b) The diagram shows a section through a road which has a sloping bank and hedge on each side. The following plant species were found growing on 10-metre lengths of the north-facing and south-facing banks.

Plant species	Number of plants	
	South-facing road bank	North-facing road bank
White deadnettle	23	0
Lesser celandine	18	8
Dandelion	8	4
Ragwort	10	0
Cow parsley	7	10
Thistle	5	1
Groundsel	15	10
Index of diversity	5.94	

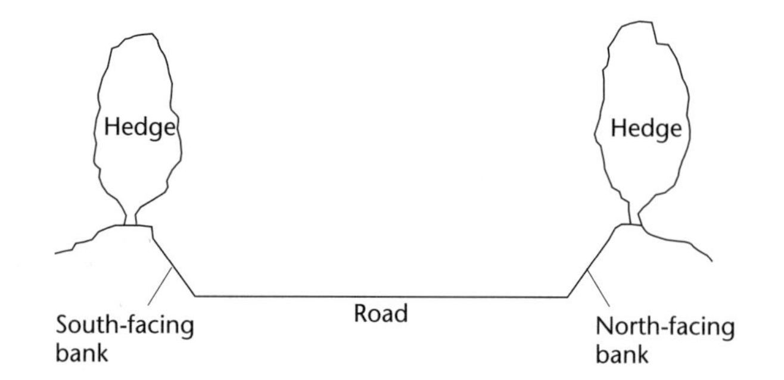

(i) Use the formula $d = \frac{N(N-1)}{\Sigma n(n-1)}$

where d = index of diversity

N = total number of organisms of all species

n = total number of organisms of a particular species

to calculate the index of diversity for the plants growing on the north-facing road bank. Show your working. *(2 marks)*

$0 + 8 + 4 + 0 + 10 + 1 + 10 = 33$

$$\frac{33 \times 32}{(8 \times 7) + (4 \times 3) + (10 \times 9) + (1 \times 0) + (10 \times 9)}$$

Index of diversity = ...

The candidate has not finished the question and has failed to calculate the index of diversity. However, she has correctly used the data from the table in the formula given in the question. This gains one mark. The correct numerical answer for the second mark is 4.26/4.3

(ii) Give **one** advantage of calculating the index of diversity rather than just recording the number of species present. *(1 mark)*

Because the index of diversity gives a more accurate answer than just counting the number of species.

'A more accurate answer' is not likely to gain a mark in any Biology Unit test. The answer here is that the index of diversity takes account of the number of individuals/population size as well as the number of species.

(iii) Suggest and explain how **one** abiotic factor might have caused differences in plant growth on the two road banks. *(2 marks)*

South facing gardens get more sunlight because the sun is to the south. Plants on the south facing road bank will get more light and so will be able to photosynthesise faster and grow more. Plants on the north facing road bank will get less light and so will need to be able to grow in dim light conditions.

This answer is fine and gains two marks for: more light; more photosynthesis. An alternative answer is: warmer on south-facing side; (warmth) allows faster metabolism.

(iv) Explain why the south-facing road bank is likely to show greater ecological stability than the north-facing road bank. *(3 marks)*

Because it has more light, there will be more plant growth. This means that there will be more food for the animals that live there. On the north facing side there will be little plant growth and so there will not be so much food for animals, especially in a bad year.

The candidate has missed the point here. A higher index of diversity indicates a less harsh environment in which abiotic factors have less influence than biotic factors.

(v) The south-facing road bank would also be expected to have a higher diversity of animals. Suggest one reason for this. *(1 mark)*

There will be a greater variety of plants which will allow a greater number of types of herbivore to feed there.

This is a fairly simple concept and the candidate gains the mark.

c) In order to estimate the population of woodlice living on the north-facing road bank, four pitfall traps were set in the ground at 2-metre intervals and left for 24 hours. All the woodlice that had fallen into the traps were marked on their underside with quick-drying paint and released back into their habitat. The next day the traps were examined again and the numbers of marked and unmarked woodlice were counted. The results are shown in the table.

	Trap number				
	1	2	3	4	
Number of woodlice marked and released	2	28	0	10	*40*
Number of woodlice in 2nd catch	0	4	0	2	*6* } *42*
Number of unmarked woodlice in 2nd catch	5	17	3	11	*36* }

(i) Use the data to estimate the woodlouse population in this area. Show your working. *(2 marks)*

Number of animals = 40 ×

Population = ...

The candidate has made a start to the calculation and has correctly identified the need to add the numbers of woodlice from each of the four traps. However, she has failed to recall the method for calculating population size and so gains no marks. Two marks are awarded for the correct answer = 280. One mark would be awarded for using the correct data/correct formula (population = (40 × 42) ÷ 6), but calculating the wrong answer.

(ii) Suggest **two** reasons why it is not possible to make a reliable estimate of the woodlouse population size from these data. *(2 marks)*

The quick drying paint might have been poisonous to woodlice but the investigators did not know it. This would mean that some of the marked woodlice might have died and so they could not be trapped again.

The question has asked for two reasons, but the candidate has given only one. Be sure you follow the instructions in the question. In this case, the candidate gets one mark for identifying that some of the marked woodlice might have died. A second reason would be any one of: sample too small / too few traps / too short a time for woodlice to mingle with rest of population; immigration / emigration; method of marking affected woodlouse behaviour.

(AQA 2002)

EXAMINATION QUESTIONS

1 a) What is meant by each of the following ecological terms?

(i) Community (ii) Population (iii) Ecosystem *(3 marks)*

b) Suggest why, in a particular ecosystem, two species do not occupy the same niche. *(2 marks)*

(AQA 2002)

5 Energy transfer

After revising this topic, you should be able to:

- explain the role of **ATP** as a source of **energy** in metabolism
- explain how **respiration** produces the ATP which is used in metabolism
- explain how **light energy** provides energy to generate ATP, which is used to synthesise organic molecules from inorganic molecules during photosynthesis
- show an understanding of **energy transfer** through biological communities

ATP as an energy source

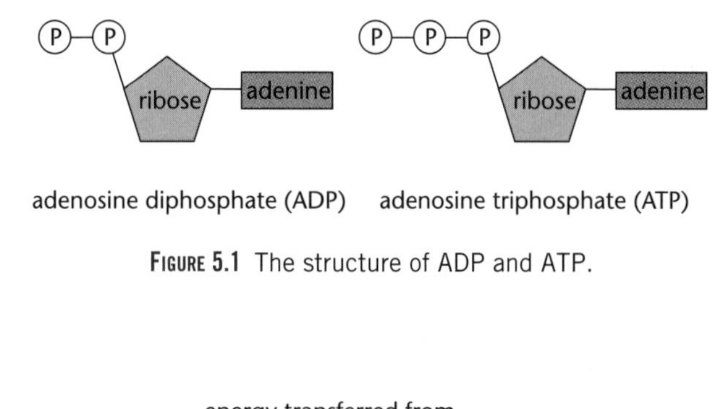

FIGURE 5.1 The structure of ADP and ATP.

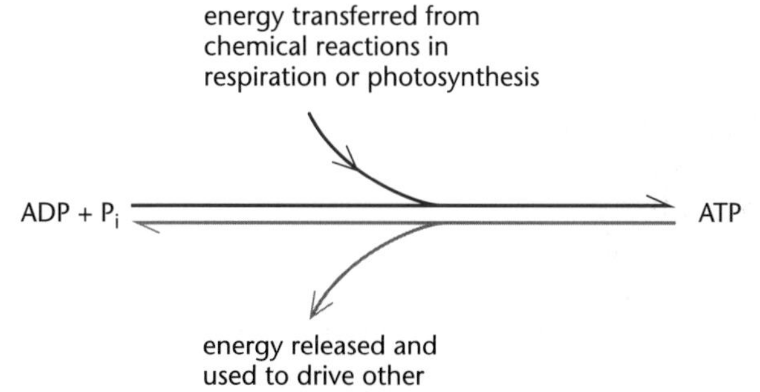

FIGURE 5.2 Synthesis of ATP.

Figure 5.1 represents the structure of **ATP** and **ADP**. In living cells, ATP is normally synthesised by **adding an inorganic phosphate group** (represented P_i) to existing molecules of ADP (Figure 5.2). This reaction requires energy, which is released during the reactions involved in respiration or photosynthesis. ATP is useful as a universal energy source for cells because:

- the breakdown of ATP to ADP is a **single-step reaction**, which makes energy immediately available for anabolic processes with cells;
- the amount of energy released by the breakdown of ATP to ADP is small enough to be used for anabolic processes **without the release of large amounts of surplus energy (heat)**.

1 The energy to synthesise ATP is released from organic molecules such as glucose. Explain why glucose is not suitable as an immediate source of energy for anabolic reactions in cells.

EXAMINER'S TIP

Always refer to energy being released. Do not refer to energy being produced, since energy cannot be created or destroyed.

Respiration

Respiration is the process by which organic molecules are **broken down, releasing energy** to synthesise ATP. It consists of the stages shown in Figure 5.3 and explained in Table 5.1.

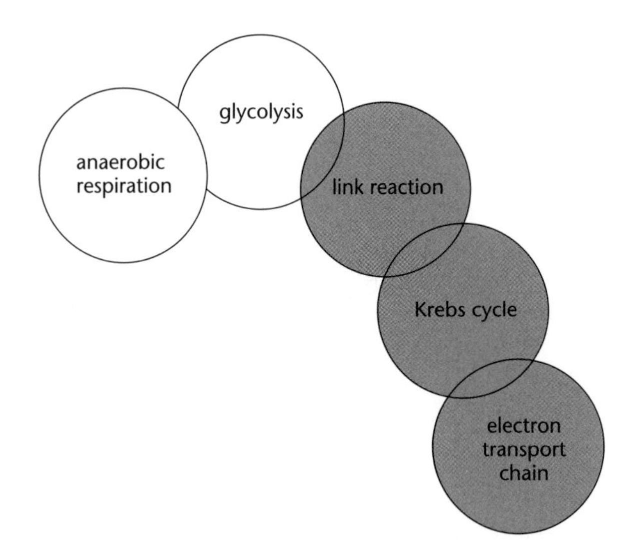

FIGURE 5.3 The main stages involved in cell respiration. The coloured circles show the stages that will only occur if oxygen is present (aerobic respiration) and which take place within mitochondria.

Stage in respiration		Explanation
Glycolysis	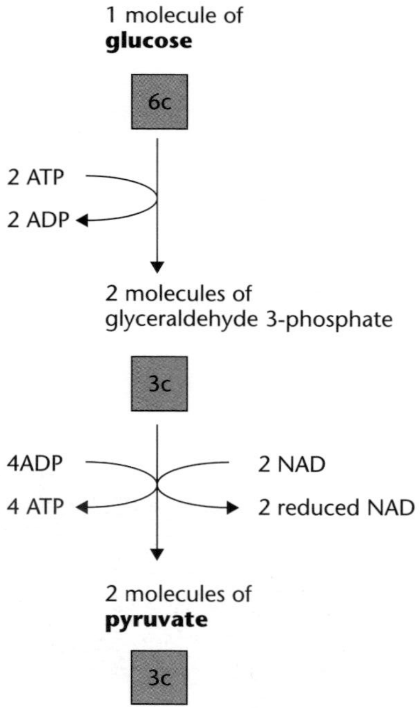 1 molecule of **glucose** 6c 2 ATP → 2 ADP 2 molecules of glyceraldehyde 3-phosphate 3c 4ADP → 4 ATP 2 NAD → 2 reduced NAD 2 molecules of **pyruvate** 3c	◆ A molecule of glucose is broken down into two 3-carbon pyruvate groups. ◆ In the first step, ATP is used both to **provide energy** to split the glucose molecule and to add phosphate groups to the 3-carbon compounds formed. ◆ In the second step, **energy is released** and used to synthesise more ATP. The coenzyme NAD (nicotinamide adenine dinucleotide) is also reduced in this step. ◆ OUTCOME: NET GAIN OF ATP, FORMATION OF REDUCED NAD AND PYRUVATE
Link reaction	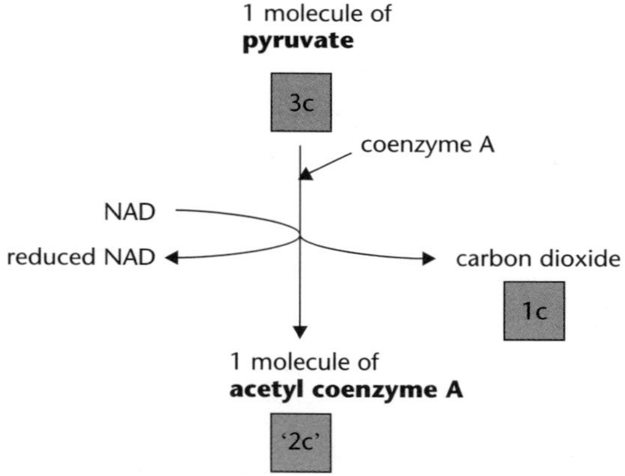 1 molecule of **pyruvate** 3c coenzyme A NAD → reduced NAD carbon dioxide 1c 1 molecule of **acetyl coenzyme A** '2c'	◆ In this reaction, pyruvate reacts with coenzyme A to produce acetylcoenzyme A. ◆ OUTCOME: FORMATION OF ACETYLCOENZYME A REDUCED NAD AND CARBON DIOXIDE

Table continues overleaf

Stage in respiration	Explanation
The Krebs cycle 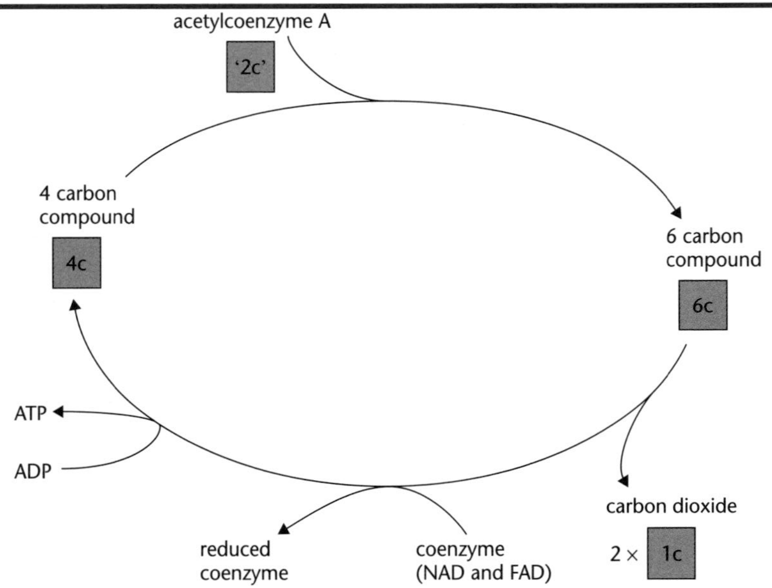 	◆ Acetycoenzyme A reacts with a 4-carbon compound to form a 6-carbon compound. In a series of oxidation-reduction reactions, this 6-carbon compound is broken down into the initial 4-carbon compound again. These reactions release carbon dioxide and also result in the synthesis of: – more ATP – more reduced NAD – reduced FAD (flavine adenine dinucleotide – another coenzyme) ◆ OUTCOME: SYNTHESIS OF ATP, REDUCED COENZYMES AND CARBON DIOXIDE
Oxidative phosphorylation involving a chain of electron carriers 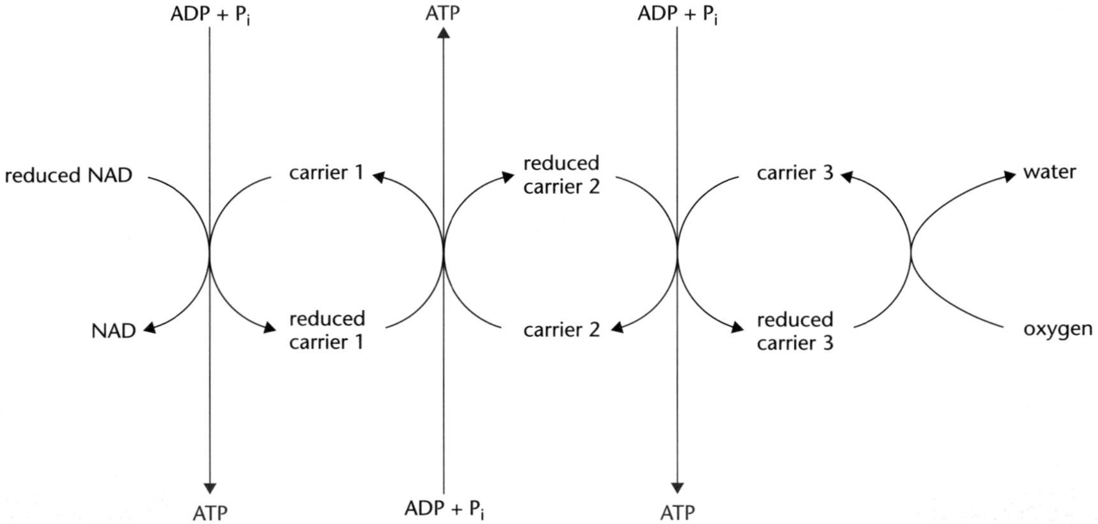 	◆ Reduced coenzyme from glycolysis, link reaction and the Krebs cycle starts a series of oxidation-reduction reactions. ◆ The final oxidation-reduction reaction in the series involves the reduction of oxygen to form water. ◆ At several steps along the series of reactions, energy is released and used to synthesise ATP from ADP and inorganic phosphate. This process is termed oxidative phosphorylation because it involves oxidation and the addition of phosphate to ADP. ◆ OUTCOME: SYNTHESIS OF ATP, OXIDATION OF COENZYMES AND REDUCTION OF OXYGEN TO FORM WATER

Table 5.1 Details of the main stages of cell respiration. Notice that the final product is ATP and that carbon dioxide and water are produced as by-products.

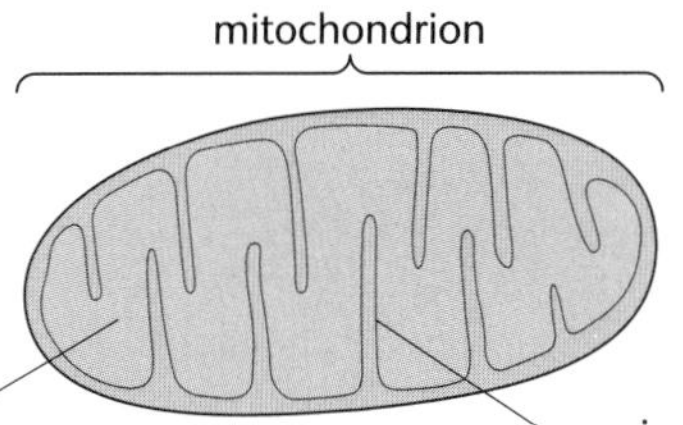

Figure 5.4 The reactions of the Krebs cycle occur within the liquid matrix of a mitochondrion. The carrier molecules of the electron transport chain lie on the inside of the inner membrane of a mitochondrion. Cristae increase the surface area of this membrane, increasing the rate of oxidation-reduction reactions.

2 What happens to a) the water and b) the carbon dioxide formed as by-products of respiration?

3 Write a word equation to represent the reaction between reduced NAD and carrier 1, shown in Table 5.1.

4 Use your knowledge of chemistry to explain why the series of oxidation-reduction reactions shown in Table 5.1 is called the electron transport chain.

Many students find the biochemical details of respiration and photosynthesis difficult to recall. Examination questions often give you some of the detail and then ask questions about it. Do ensure that you:

- recall the main products of a process
- are able to interpret a given biochemical pathway.

ANAEROBIC RESPIRATION

Without oxygen the final reaction of the electron transport chain cannot occur. As a result, the reactions of the electron transport chain and of the Krebs cycle stop. In these circumstances, pyruvate is converted to the other compounds shown in Figure 5.5. These pathways are termed **anaerobic respiration** and occur in the **cytoplasm of cells**, i.e. not within the mitochondria.

Anaerobic respiration enables organisms to survive in environments with **low concentrations of oxygen**. However, the amount of ATP produced by anaerobic respiration is **much less** than that produced by aerobic respiration.

5 Yeast is used in the brewing industry. Suggest why yeast cells are allowed to respire aerobically in the early stages of brewing but later are made to respire anaerobically.

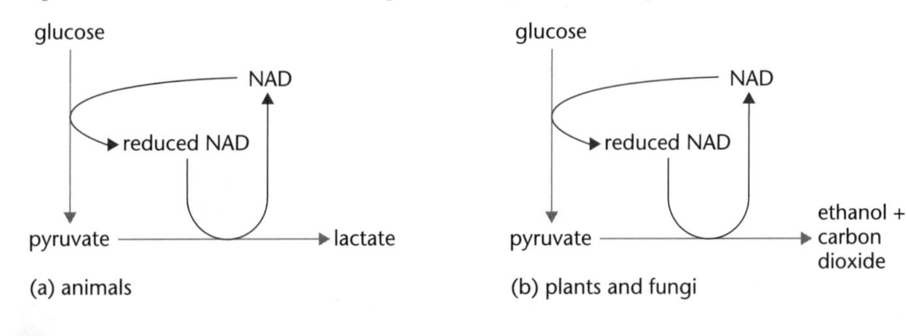

Figure 5.5 In the absence of oxygen, anaerobic respiration takes place. The biochemical pathways of anaerobic respiration are different in animals and plants.

The breakdown of ATP is always linked to other reactions, providing them with the energy they need. Examples of linked reactions are given in this chapter, e.g. in glycolysis and the light-independent reactions of photosynthesis. Make sure you relate the use of ATP to processes in other modules: the use of ATP would make a good synoptic question!

RESPIRATORY SUBSTRATES AND RESPIRATORY QUOTIENT (RQ)

In the reactions above, the respiratory substrate was glucose. However, other organic compounds can be used instead. We can investigate which **respiratory substrate** is being used by **comparing** the **volume of carbon dioxide** that is produced and the **volume of oxygen** that is used by living organisms in a given period of time. The equation below is often used to represent the respiration of glucose.

$$C_6H_{12}O_6 + 6O_2 \longrightarrow 6CO_2 + 6H_2O$$

Notice that the number of molecules of carbon dioxide produced is the **same** as the number of molecules of oxygen used. This would also be true of the **volumes** of these gases. This ratio is known as the **respiratory quotient**, or RQ. It is calculated as:

$$RQ = \frac{\text{volume of carbon dioxide produced}}{\text{volume of oxygen used}} \text{ per unit time}$$

Table 5.2 shows values of the RQ derived from equations summarising the respiration of carbohydrates, proteins and triglycerides. Often cells respire **more than one** respiratory substrate at once. An RQ of 0.9 could mean that an organism is respiring only proteins, or that it is respiring a mixture of carbohydrates and triglycerides.

Respiratory substrate	RQ
Carbohydrate	1.0
Protein	0.9
Triglyceride	0.7

TABLE 5.2 The RQ values of common respiratory substrates.

? 6 **a) What is meant by the term respiratory quotient?**
b) What would an RQ value of 0.95 suggest about the respiratory substrate?

Photosynthesis

Photosynthesis involves:

- **trapping of light energy** to form ATP from ADP and P_i
- **fixation of carbon dioxide**, i.e. carbon dioxide is converted into organic compounds.

These steps occur in a different series of reactions, shown in Figure 5.6.

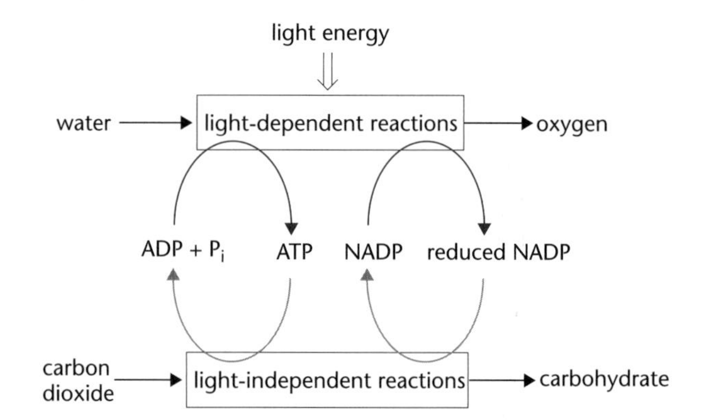

FIGURE 5.6 Photosynthesis involves two series of reactions: the light-dependent reactions and the light-independent reactions. ATP formed in the light-dependent reactions is used in the light-independent reactions. A coenzyme, NADP (nicotinamide adenine dinucleotide phosphate) is reduced in the light-dependent reactions and this reduced coenzyme is used in the light-independent reactions.

LIGHT-DEPENDENT REACTIONS

In these reactions:

- **electrons** in chlorophyll molecules are **excited** by light energy (in other words, their energy level is raised and they leave the molecule)
- **energy** from these excited electrons is **used** to synthesise ATP from ADP and P_i and to reduce a coenzyme, NADP
- **water is split** into protons, electrons and oxygen. The oxygen is released as a waste product.

If you are unfamiliar with chemistry, do not lose track of these three basic concepts. Table 5.3 gives more detail about these processes.

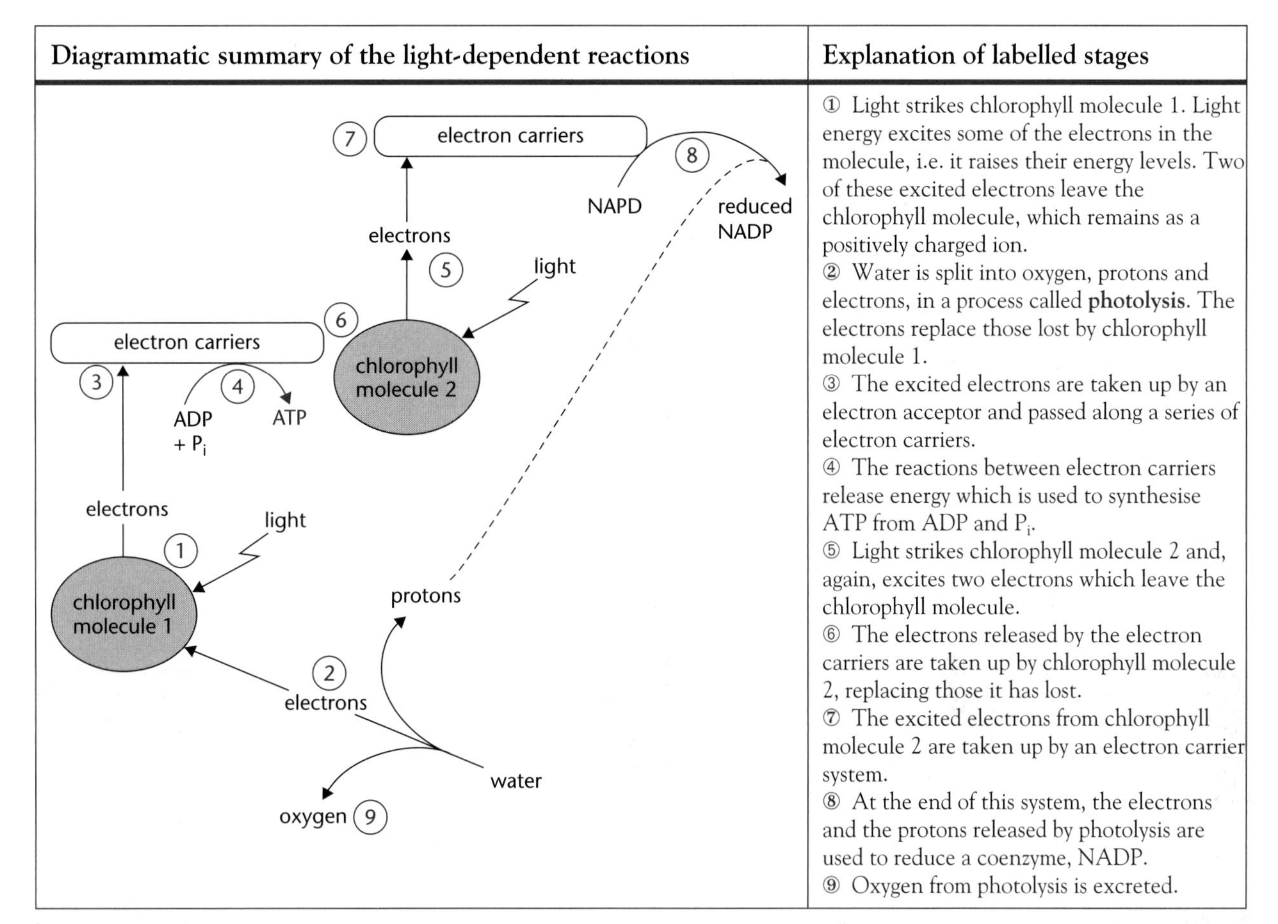

Diagrammatic summary of the light-dependent reactions	Explanation of labelled stages
	① Light strikes chlorophyll molecule 1. Light energy excites some of the electrons in the molecule, i.e. it raises their energy levels. Two of these excited electrons leave the chlorophyll molecule, which remains as a positively charged ion. ② Water is split into oxygen, protons and electrons, in a process called **photolysis**. The electrons replace those lost by chlorophyll molecule 1. ③ The excited electrons are taken up by an electron acceptor and passed along a series of electron carriers. ④ The reactions between electron carriers release energy which is used to synthesise ATP from ADP and P_i. ⑤ Light strikes chlorophyll molecule 2 and, again, excites two electrons which leave the chlorophyll molecule. ⑥ The electrons released by the electron carriers are taken up by chlorophyll molecule 2, replacing those it has lost. ⑦ The excited electrons from chlorophyll molecule 2 are taken up by an electron carrier system. ⑧ At the end of this system, the electrons and the protons released by photolysis are used to reduce a coenzyme, NADP. ⑨ Oxygen from photolysis is excreted.

Table 5.3 Details of the light-dependent reactions.

? 7 Name two products of the light-dependent reactions that are used in the light-independent reactions.

LIGHT-INDEPENDENT REACTIONS

In these reactions:

- a 4-carbon compound, ribulose bisphosphate (RuBP), **reacts with carbon dioxide** to form two 3-carbon molecules of glycerate 3-phosphate (GP)
- ATP and reduced NADP from the light-dependent reactions are used to **reduce GP** to triose phosphate (TP)
- some triose phosphate is **converted** into useful carbohydrates, amino acids and triglycerides
- some triose phosphate is used to **regenerate** RuBP in the Calvin cycle.

If you are unfamiliar with chemistry, do not lose track of these four basic concepts. Figure 5.7 gives more detail about these processes.

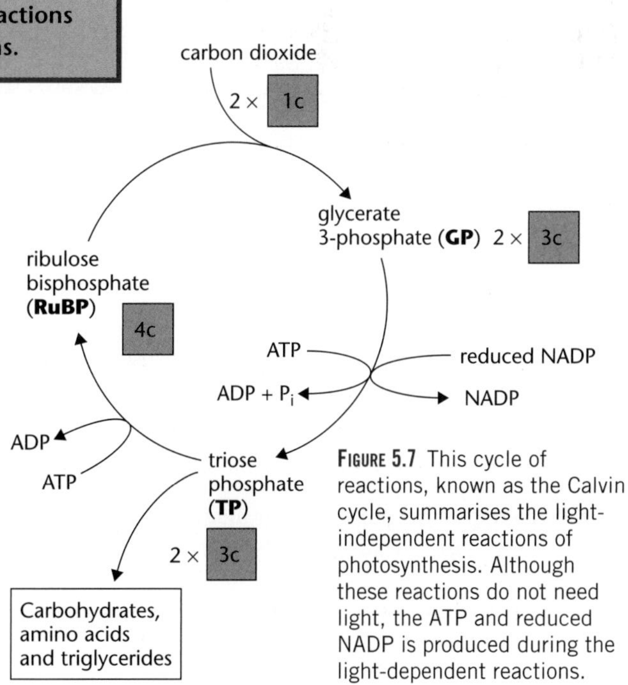

Figure 5.7 This cycle of reactions, known as the Calvin cycle, summarises the light-independent reactions of photosynthesis. Although these reactions do not need light, the ATP and reduced NADP is produced during the light-dependent reactions.

Both the light-dependent and the light-independent reactions occur within **chloroplasts**. Figure 5.8 shows the structure of a chloroplast. Chlorophyll molecules, together with other light-trapping pigments, are located within its internal membranes – or **thylakoids**. In some places, the thylakoids are arranged in stacks, called **grana**, which is where the light-dependent reactions occur. The fluid **stroma** contains the chemicals involved in the Calvin cycle.

Exam questions often contain part of a biochemical process, such as the Calvin cycle, and ask you to identify the number of carbon atoms in identified molecules. These are easy marks – make sure you are prepared to gain them.

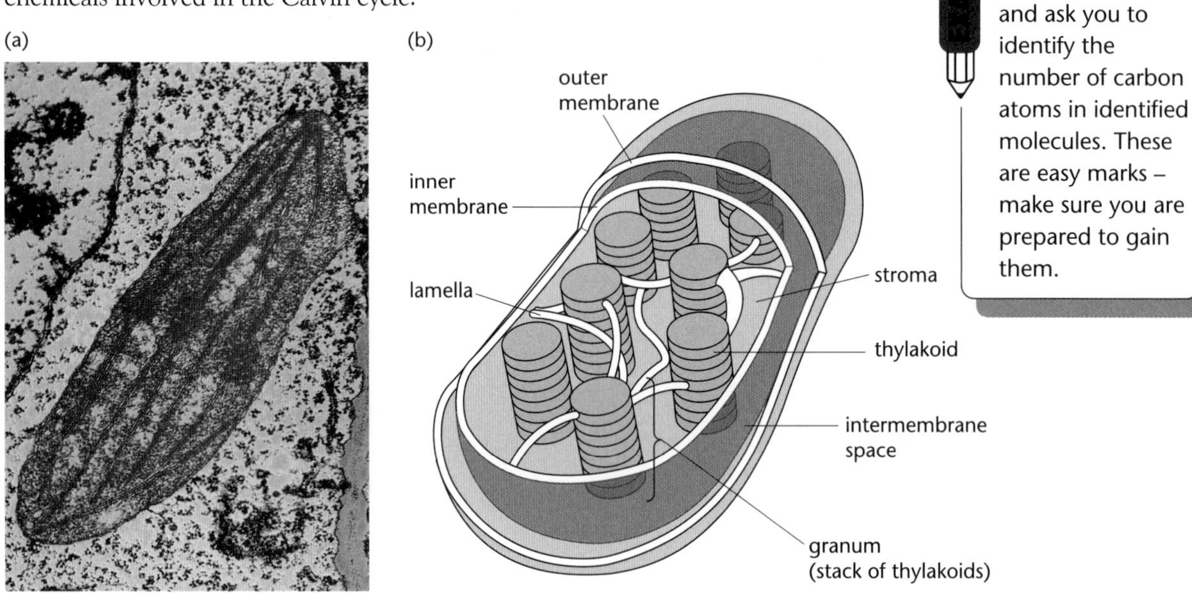

Figure 5.8 a) A transmission electron micrograph of a single chloroplast. b) A diagrammatic representation of the internal structure of a chloroplast.

8 In an investigation into the light-independent reactions, a photosynthesising plant was supplied with air containing radioactively labelled carbon dioxide ($^{14}CO_2$).
a) In which two compounds would you first expect to find the radioactivity?
b) Name one process by which you could separate and identify these compounds from a plant cell.

Energy transfer in biological communities

Neither respiration nor photosynthesis is efficient at transferring energy.

- In respiration and photosynthesis, energy is lost as heat energy.
- In photosynthesis, light energy is not used because it is the wrong wavelength, is reflected from a plant's surface, or falls on parts of the plant that lack chloroplasts.

In biological communities, populations obtain their energy in different ways.

- **Producers** are able to **manufacture** their own energy-containing nutrients. Plants are producers; they manufacture their own energy-containing compounds during photosynthesis.
- **Consumers** obtain their energy-containing nutrients by **consuming** other organisms. Many bacteria and protoctists and all fungi and animals are consumers.

Figure 5.9 shows how energy is **transferred** between different populations in a community. The feeding pattern of each population is called its **trophic level** and includes producer, primary consumer (eats producers), secondary consumer (eats primary consumers) and decomposer. A **food chain** shows individual populations from each trophic level, e.g. bean plant → bean aphid → two-spot ladybird.

The data in Figure 5.9 show that energy capture and its transfer between trophic levels is **inefficient**: most of the available energy is lost as heat. This loss of energy explains the shape of the **pyramid of energy** shown in Figure 5.10.

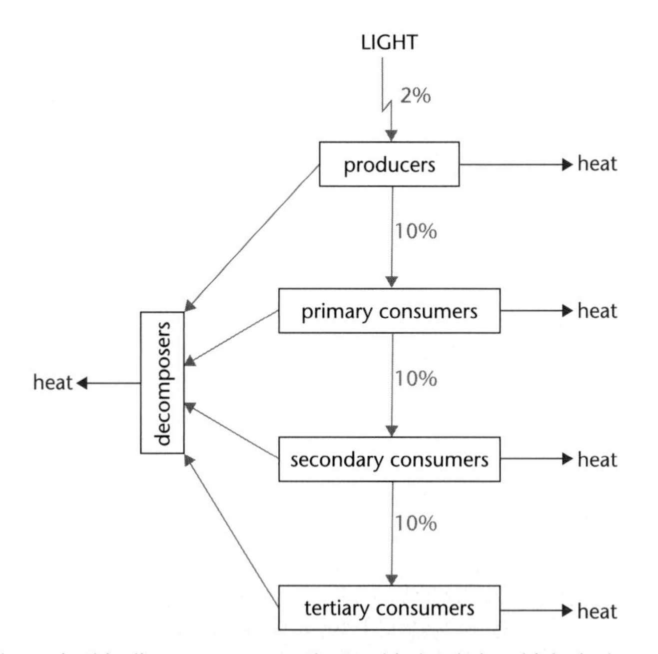

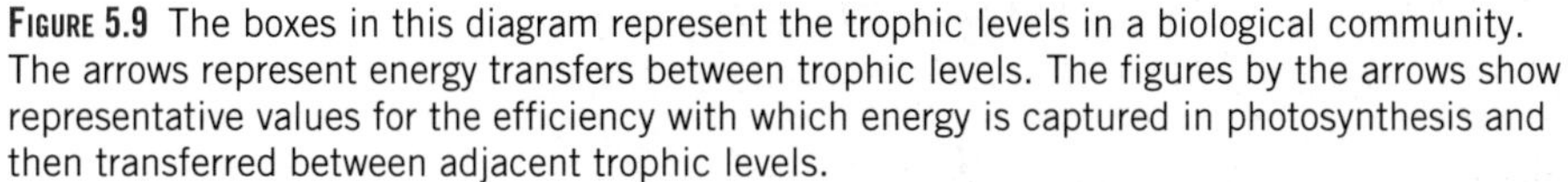
FIGURE 5.9 The boxes in this diagram represent the trophic levels in a biological community. The arrows represent energy transfers between trophic levels. The figures by the arrows show representative values for the efficiency with which energy is captured in photosynthesis and then transferred between adjacent trophic levels.

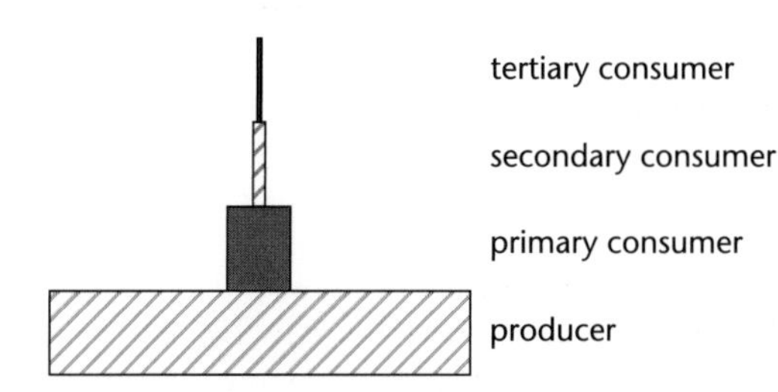

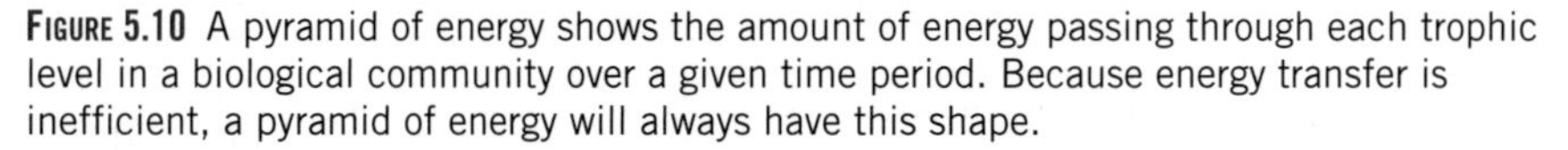
FIGURE 5.10 A pyramid of energy shows the amount of energy passing through each trophic level in a biological community over a given time period. Because energy transfer is inefficient, a pyramid of energy will always have this shape.

We can also construct **pyramids of number** and **pyramids of biomass** to represent, respectively, the **number of organisms** and the **total dry mass of organisms** in each trophic level. Although they usually have a similar appearance to the pyramid of energy, Table 5.4 shows that other shapes can occur.

	Typical pattern	Unusual patterns
Pyramid of number	e.g. there are more grass plants than rabbits and more rabbits than foxes	e.g. a small number of oak trees support vast numbers of insects, which are fed on by a smaller number of birds
Pyramid of biomass	e.g. there is a greater biomass of grass plants than of rabbits and a greater mass of rabbits than of foxes that feed on them	e.g. at any one time, a small but rapidly reproducing population of phytoplankton can support a larger population of primary consumers

TABLE 5.4 Investigations of biological communities can provide data which can be used to produce pyramids of number and pyramids of biomass.

9 **For the food chain, grass → rabbit → fox: a) name the primary consumer, b) describe and explain the change in biomass along the food chain.**

WORKED EXAM QUESTIONS

1 The diagram gives an outline of the process of aerobic respiration.

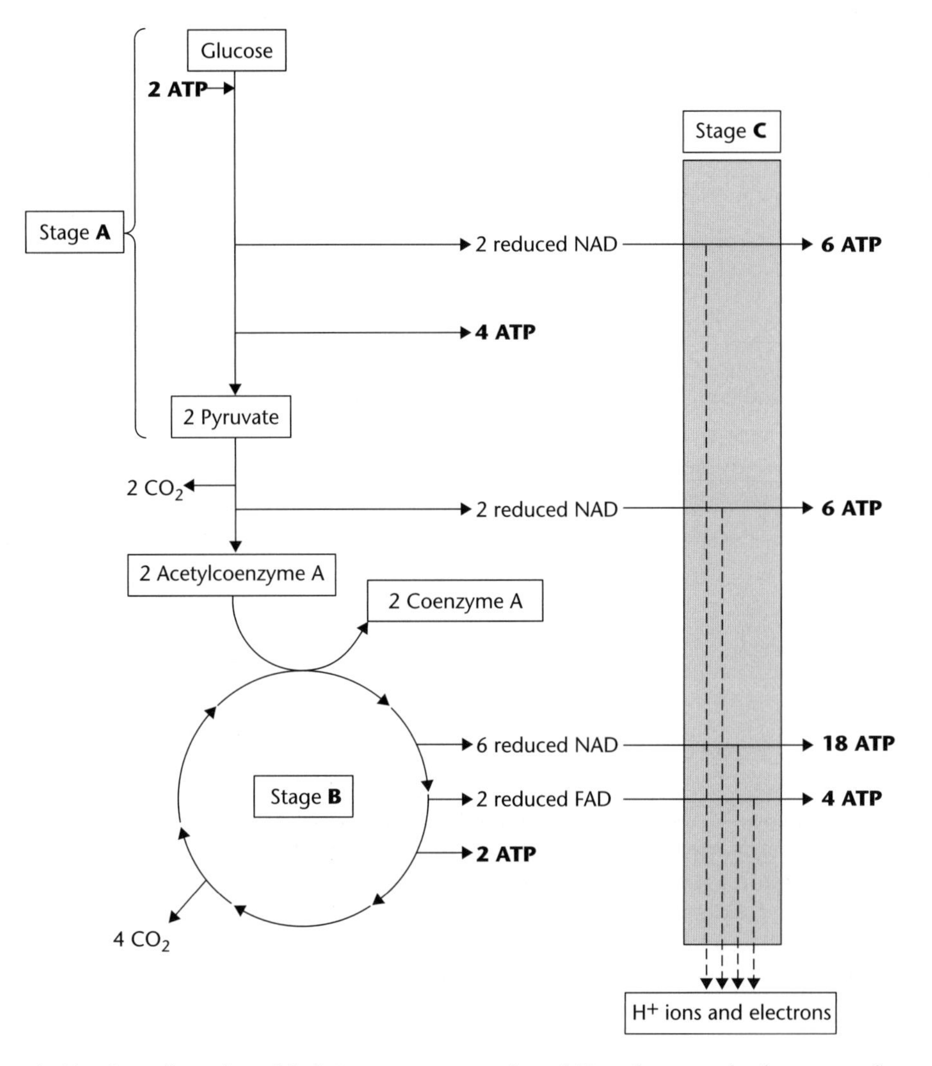

a) (i) Complete the table by naming stages **A** and **B** and giving the location of each stage in a cell such as a liver cell. *(2 marks)*

Stage	Name of stage	Location in cell
A	*glycolysis*	*cytoplasm*
B	*Kreb cycle*	*mitochondria*

This question is marked along the rows of the table. Both answers about stage A must be correct to gain a mark and the same for stage B. In this case, the candidate gains both marks. The examiner has not penalised the candidate for stating 'Kreb' rather than 'Krebs'.

(ii) How many carbon atoms are there in each pyruvate ion? *(1 mark)*

3

> The candidate is correct and gains the mark. We do not know whether the candidate recalled this, or whether she used information in the diagram: working out that dividing a molecule of glucose into two pyruvates would mean 6 carbon atoms divided by 2 = 3 each.

(iii) What happens to the H^+ ions and electrons released at stage **C**? *(2 marks)*

They are passed down the electron transport chain. H^+ ions combine with FAD and electrons are used to produce ATP by ADP + P_i.

> The diagram in the question is not a usual way of representing respiration and this candidate has failed to interpret it correctly. She did not realise that stage C is the electron transport chain and that the formation of ATP by this chain is shown on the diagram. In fact, the hydrogen ions and electrons shown are at the very end of the electron transport chain. The correct answer was: 'combine with oxygen; to form water'.

b) In aerobic conditions, ATP is produced by substrate-level phosphorylation and by oxidative phosphorylation. Use information in the diagram to find the net yield of molecules of ATP per molecule of glucose by:

(i) substrate-level phosphorylation **6 − 2 = 4**

(ii) oxidative phosphorylation **34** *(2 marks)*

> To answer this question correctly, the candidate needed to recall that oxidative phosphorylation occurs during the electron transport chain (stage C) and then either recall or deduce that substrate-level phosphorylation must relate to ATP production in stages A and B. Having realised this, the question becomes a simple addition: two ATPs are used in stage A, after which a further four ATPs are produced in stage A and two are produced in stage B. This gives a net total of 4. In stage C, you needed only to add the numbers of the right-hand side of the diagram (6 + 6 + 18 + 4) to get the correct answer of 34 ATPs.

c) (i) One mole of glucose releases 2880 kJ of energy when burned completely in oxygen. Hydrolysis of one mole of ATP to ADP and phosphate releases 31 kJ of energy. Use your answers from part b) to calculate the percentage efficiency of energy transfer from glucose to ATP by aerobic respiration. Show your working. *(2 marks)*

34 + 4 = 38 molecules of ATP from above
they contain 38 × 31 kJ of energy
1 molecule of glucose contains 2880 kJ of energy

$$\text{Efficiency} = \frac{38 \times 31}{2880} \times 100\% = 40.9$$

Answer **41%**

Notice that the question asks you to use your answer to part b). You might worry that if you got part b) wrong, you will get this wrong too. Examiners ensure this does not happen. Whatever your answer to part b)(i) and (ii), if you had correctly used them you could have gained the two marks. The correct answer of 41% gains two marks, even without showing your working. However, the mark scheme is:

$$\frac{[\text{answer from b)(i)} + \text{(ii)} \times 31]}{2880} \times 100 \text{ ; correctly calculated answer;}$$

It is a good idea to show your working. Even if you make a mistake in your calculation, you can gain one mark for the correct procedure.

(ii) What happens to the energy which is **not** transferred to ATP? *(1 mark)*

It remains in the bonds in the compounds.

This is a recall question, which you can attempt even if you were troubled by the calculation. This candidate got the answer wrong. The correct answer is: 'lost as heat'.

(iii) Explain why ATP is better than glucose as an immediate source of energy for cell metabolism. *(2 marks)*

The energy is released faster and is better obtainable i.e. there are ATP molecules readily available in mitochondria whilst glucose is not. If stored, glucose would be in a form of glycogen and the process of glucogenesis would have to occur.

As often happens, this candidate has wandered off the main point here. Her answer about glycogen and the incorrect reference to glucogenesis (she means glycogenolysis) are both irrelevant.
She gains a mark for stating that 'energy release is faster' ('released in a single step' would be a suitable alternative answer) but fails to gain the second mark for: 'energy released in small/manageable quantities'.

(iv) Give three uses of energy from ATP in a liver cell. *(3 marks)*

1 ***used for synthesis of macromolecules***
2 ***active transport of glucose into the liver cell in times of high blood glucose levels***
3 ***used for ~~mat~~***

The mark scheme is fairly long for this question. A mark would be awarded for any of the following, to a maximum of three marks.

Active transport; phagocytosis; synthesis of glycogen; synthesis of protein/enzyme; synthesis of DNA/RNA; synthesis of lipid/cholesterol; synthesis of urea; bile production; cell division.

So why did this candidate score zero on this question? Look at her first answer. She has got the idea of energy being used for synthesis. However, given that the question specified a liver cell, she was required to name an appropriate macromolecule to gain the mark. Her failure to gain a mark for her second answer might puzzle you. She has written 'active transport', which is in the mark scheme. Had she stopped there, she would have gained the mark, but read on. She has written active transport (correct) of glucose (wrong). Glucose is not actively transported into cells. As a consequence, the mark she gains for 'active transport' is cancelled by the incorrect identification of glucose.

(AQA 2002)

EXAMINATION QUESTIONS

1 In an investigation, leaf cells were supplied with $^{14}CO_2$, carbon dioxide labelled with a radioactive isotope of carbon. These cells were kept in the light and allowed to photosynthesise. After a period of time, the light was switched off and the cells were left in the dark. The graph shows the concentration of radioactively-labelled glycerate 3-phosphate (GP) over the course of the investigation.

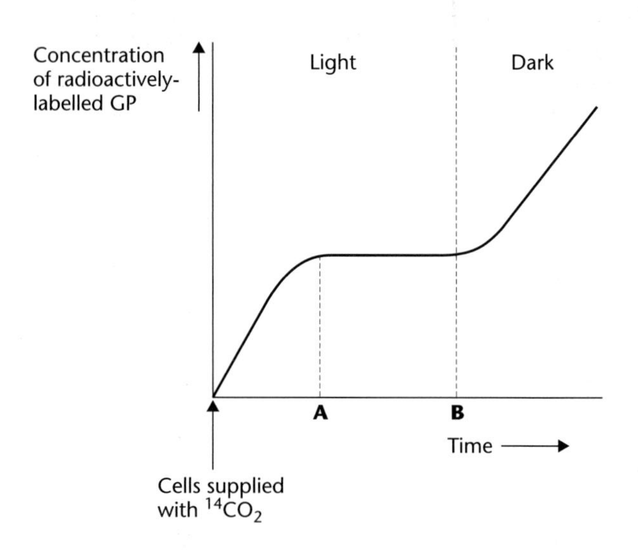

a) (i) Describe how GP is formed from carbon dioxide in photosynthesis. *(2 marks)*

(ii) In this investigation, the $^{14}CO_2$ was supplied in excess. Explain why the concentration of radioactively-labelled GP remained the same between times **A** and **B** on the graph. *(1 mark)*

b) Explain the change in the concentration of radioactive GP after the light was switched off. *(3 marks)*

The table shows some results from an investigation of the concentration of carbon dioxide in samples of air taken from among the leaves in a potato crop.

Date	Mean carbon dioxide concentration in parts per million between	
	8 pm and 4 am	8 am and 4 pm
10 July	328	309
20 July	328	299
30 July	326	284
10 Aug	322	282

c) (i) The figures in columns 2 and 3 of the table were calculated from readings obtained at different times of the day. Explain why the figures in column 3 are lower than those in column 2. *(2 marks)*

(ii) How would you expect the mean carbon dioxide concentration between 8 am and 4 pm to have differed if the air samples had been collected at soil level? Give a reason for your answer. *(2 marks)*

d) Suggest why, in this investigation, the investigators recorded the wind speed. *(2 marks)*

e) Some of the leaves from this crop die and fall to the ground. Describe how the carbon contained in the dead leaves becomes available and can be taken up by plants. *(3 marks)*

(AQA 2003)

6 Decomposition and recycling

After revising this topic, you should be able to:

- understand the general principles involved in the **cycling of nutrients**
- describe the basic details of the **carbon cycle**
- describe the part played by **microorganisms** in the cycling of carbon
- describe the basic details of the **nitrogen cycle**
- describe the part played by **microorganisms** in the cycling of nitrogen
- understand how the cycling of carbon and nitrogen is similar to the cycling of other nutrients
- understand the importance of respiration and photosynthesis in giving rise to **short-term fluctuations** and **long-term global balance** of oxygen and carbon dioxide
- describe the process of **deforestation** for increasing agricultural land and the effect of deforestation on diversity and on carbon and nitrogen cycling
- understand that conservation of forests allows sustainable provision of resources

Nutrient cycles

All living organisms require a variety of nutrients for healthy growth.

- **Plants** obtain these elements in one of two ways:
  - as **ions** from the **soil** e.g. phosphates, nitrates
  - as **molecules** from the **air** e.g. carbon dioxide, if the element is carbon
- **Animals** and **other consumers** obtain these elements by:
  - **eating plants** which contain the element
  - **eating animals** which contain the element
- The nutrients, which are taken into the organism, are used in chemical reactions and are **converted into organic substances**, which form cells and tissues.

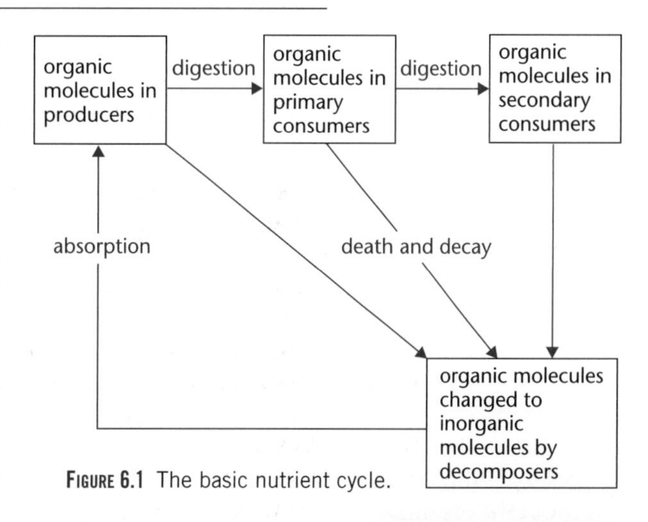

FIGURE 6.1 The basic nutrient cycle.

- Elements can be passed from one organism to another in a **food chain**.
- Organisms, if not eaten, **die and decompose**.
- Their elements, which are locked up in cells and tissues as organic molecules, are broken down by microorganisms, such as bacteria.
- The microorganisms will absorb some of the nutrients but the rest are **released as inorganic substances** e.g. ions (ammonium) or molecules (carbon dioxide), which can be taken up by the plants.

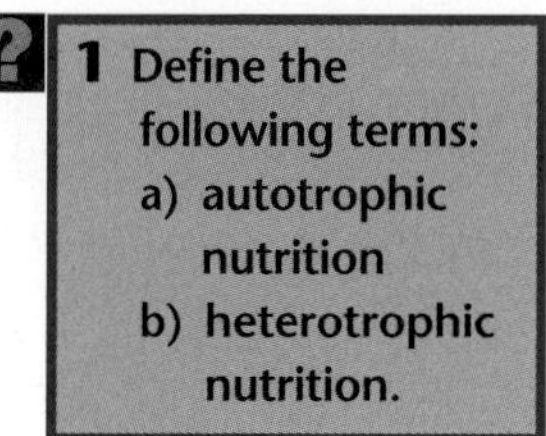

1 Define the following terms:

a) autotrophic nutrition

b) heterotrophic nutrition.

The carbon cycle

Carbon is a component of all the major biological molecules:

- Carbohydrates
- Lipids
- Proteins
- Nucleic acids (DNA, RNA)

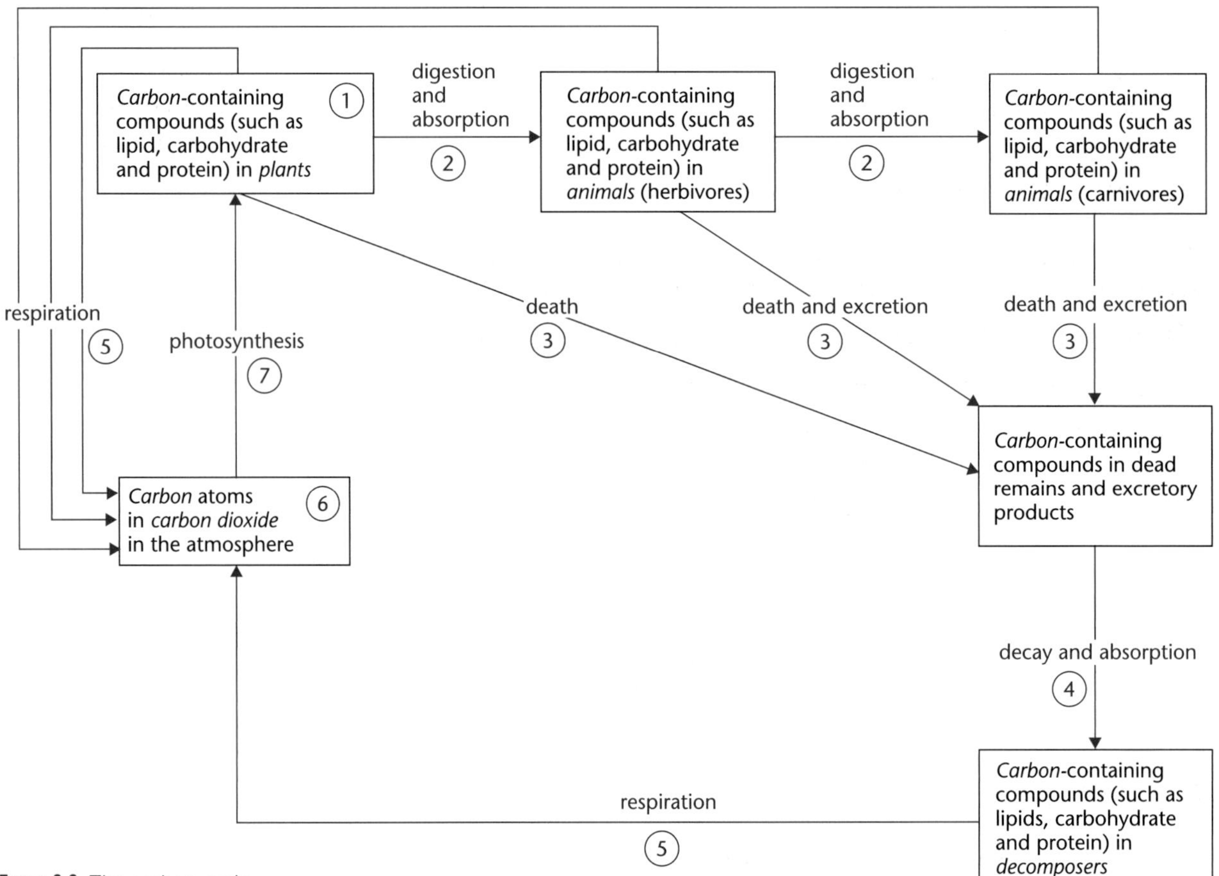

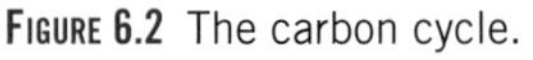
FIGURE 6.2 The carbon cycle.

1. Plants take up carbon in the form of **carbon dioxide**. It diffuses into the cells and photosynthesis converts it to carbohydrates and other organic compounds.
2. Feeding passes the carbon-containing compounds along the **food chain** to the next trophic level, where they become part of the cells and tissues of that organism. Primary consumers eat the plants and secondary consumers eat the primary consumers.
3. Any of the organisms in the food chain can die. Undigested food contains carbon compounds and passes out of animal bodies.
4. **Decomposers**, mainly bacteria, secrete enzymes which break down the large organic molecules. Some of the products of this breakdown are used by the bacteria for respiration.
5. All the organisms in the food chain respire.
6. Respiration releases carbon dioxide.
7. The carbon dioxide can be used by plants for photosynthesis.

Respiration, photosynthesis and carbon dioxide

The levels of carbon dioxide fluctuate daily. In daylight plants respire and photosynthesise; at night they only respire.

- Thus, there is more carbon dioxide in the air surrounding plants at night and less during the day.

The levels of carbon dioxide fluctuate during the year. In winter, due to cooler temperatures, shorter day length and loss of leaves, many plants reduce the amount of photosynthesis taking place. In summer, this is reversed.

- Thus, less carbon dioxide is used during the winter, and the level of carbon dioxide rises. In summer, carbon dioxide levels fall.

2 The volume of oxygen produced in a given time period is used to calculate the rate of photosynthesis. Suggest why, at low light intensities, no oxygen is released from the plant.

Human activities and the carbon cycle

We have concentrated on the fact that carbon is cycled, but we tend to forget that a great deal of carbon is locked up in **fossil fuels** such as coal and oil.

- Burning these fossil fuels releases carbon dioxide into the atmosphere.

The quickest way to clear a rainforest is to cut it down and burn the trees.

- The trees contain a lot of carbon compounds locked up in their tissues.
- Burning these releases carbon dioxide into the atmosphere.

The result of human activity is that:

- there has been a **steady increase** in the levels of carbon dioxide in the atmosphere
- this has been linked to **global warming**
  - Carbon dioxide is one of the many '**greenhouse gases**', which prevent too much of the earth's heat from escaping into space. The extra carbon dioxide produced reflects even more heat back to earth so that the earth becomes warmer.

Nitrogen cycle

Nitrogen is a component of many of the major biological molecules:

- Proteins (amino acids)
- Nucleic acids (DNA, RNA)
- ATP

1. Plants take up **nitrate ions** from the soil.
2. They are used by the plants to produce nitrogen-containing compounds such as **protein**.
3. Primary consumers eat the plants and the protein is digested to form **amino acids**. These are absorbed in the gut and are used to make up the tissues of the primary consumer.
4. This process is repeated when primary consumers are eaten by secondary consumers. In this way, nitrogen is passed from one trophic level to the next as nitrogen-containing compounds.
5. Plants and animals die; their bodies contain nitrogen-containing compounds. Animals also excrete nitrogenous substances e.g. urea. These compounds are available to decomposers

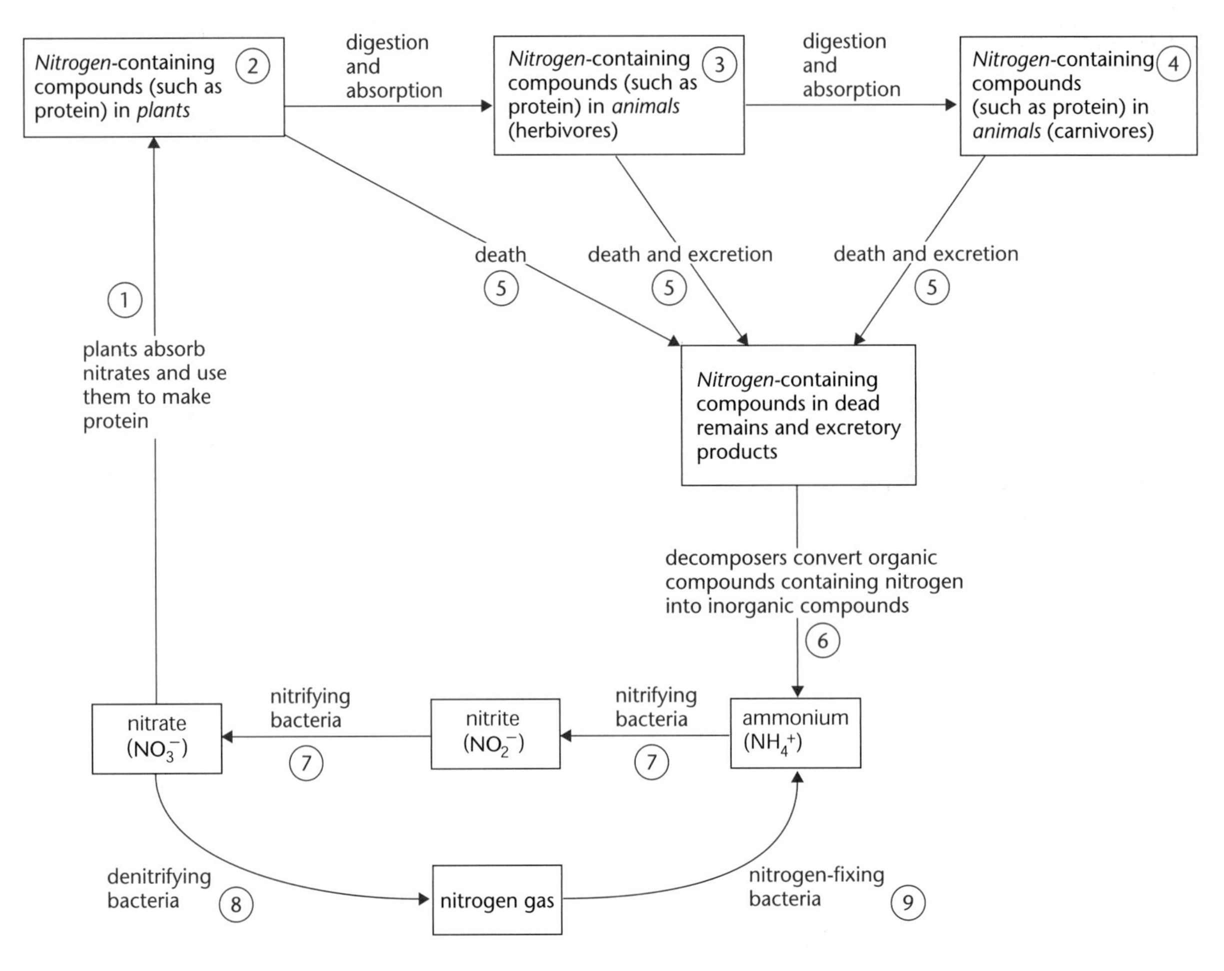

Figure 6.3 The nitrogen cycle.

6 The nitrogen-containing compounds are digested by decomposers, releasing **ammonium ions** into the soil.

7 Nitrifying bacteria convert the ammonium ions to **nitrites** and then to **nitrates**.

In addition there are two more processes that need to be considered.

8 In **anaerobic** conditions, **denitrifying bacteria** are found in large numbers.

- These bacteria use nitrates as a supply of oxygen for respiration.
- The result is that the nitrates are **reduced to nitrogen gas**, which is released into the atmosphere.
- Less nitrates are available for plants.

9 Nitrogen gas can be made available to plants by **nitrogen fixation**.

- Nitrogen-fixing bacteria live free in soil, or are associated with the roots of **leguminous plants** such as peas and beans.
- These bacteria convert nitrogen gas into ammonium compounds.
- The **enzyme** which achieves this conversion is called nitrogenase, and only works in aerobic conditions.
- The process requires much energy.

? 3 What is the benefit of the process of nitrification to nitrifying bacteria?

Deforestation

A tropical rainforest is one of the most species-rich ecosystems in the world.

Forests are cleared for many reasons, such as:

- for agricultural use
- for mining
- for timber to make furniture or as a building material
- to make room for roads and towns.

When a forest is cleared to make way for crops, a lot of the nitrogen in the soil is lost when the trees are burnt but the ash that remains is comparatively rich in the nutrients required by crops. So though you may get a good crop for the first few years the amount quickly declines. This is due to:

- the nutrients being **removed** with the crop as it is harvested: they are not replaced in the soil, as the crop is not left to rot.
- the nutrients **leaching** out of the soil by rain.

WORKED EXAM QUESTIONS

1 Read the following passage.

There will always be questions in your exam paper, which need more than a single word or a sentence as an answer. This topic lends itself to this type of extended writing. Always plan your answer. Think logically and try not to miss out any of the steps you have learnt. It is very easy to go around in circles and to repeat yourself; try to avoid that. This topic is also referred to often in the newspaper and on TV: sometimes in a very unscientific way. Be sure that your answers use the material you have been taught, and are of A-level quality.

Deforestation, particularly in tropical rain forests, has proceeded at an alarming rate over the last 60 years. In addition to local decreases in biodiversity, there have been other, global, effects, especially on climate. Deforestation has been paralleled by an enormous rise in the burning of fossil fuels. These two factors have been important contributors to rising levels of
5 carbon dioxide in the earth's atmosphere and increases in global temperatures.

Recent discussions on what do do about global warming have produced two new ideas. One is that farmers should plant trees on their land to act as carbon 'sinks'. These would offset increased carbon dioxide emissions by industry.

The second idea is that farmers should reduce the amount of ploughing they do. Ploughing
10 allows air to enter the soil and helps with the recycling of both carbon and nitrogen. A reduction in ploughing would cut the oxidation of organic matter being stored in soils, which would then act as another carbon 'sink'.

Use information from the passage and your own knowledge to answer the following questions.

a) Explain how deforestation could lead to decreases in biodiversity (line 2). *(3 marks)*

Deforestation means the removal of large numbers of trees. That in turn reduces the places that animals can use as homes. Trees are also food for herbivores. As they have less food some of them will die. Carnivores will have fewer herbivores to eat and they will die too. Eventually there will be fewer animals of any type living in the area and the diversity will be lower.

Loss of food is a valid point, but this candidate has made a meal of it by tracing the problem through the whole food chain. Removal of trees will remove habitats or nest sites or shelter (all valid reasons for a reduction of diversity) but the use of the term 'home' is not good enough for an A-level candidate. Try to think of the effect on the whole area. Factors such as exposure of the soil will lead to erosion and the removal of ions, reducing the likelihood of colonisation by other plants. The levels of light, temperature and humidity will also be affected and these will, in turn, cause organisms to die or to leave the area.

b) Explain how trees can act as carbon 'sinks' (line 7). *(3 marks)*

Trees are carbon-sinks because they remove carbon. Trees remove carbon from the air during the process of photosynthesis. Carbon reacts with water that the trees take up from the soil to make starch which is stored in the leaves, stems and roots of the tree.

The basic ideas are here for two marks. The candidate has mentioned that photosynthesis will result in the production of organic molecules, which 'fix' the element carbon within the tree. He has made one error however, which will always cost you a mark. Remember that carbon is an element and is always a solid. It exists as graphite or diamond. Trees cannot use this element: they use CARBON DIOXIDE in the process of photosynthesis, not carbon.

c) (i) Explain how a reduction in the amount of ploughing would lead to more carbon being stored in the soil (lines 9–12). *(4 marks)*

Ploughing stirs up the soil and mixes up the organic material the soil contains. Worms and other animals living in the soil are able to rot this material and release the carbon it contains so that less carbon would be in the soil. If there were less ploughing the opposite would happen. Worms would not break down the organic material and more carbon would remain in the organic material in the soil.

Not a good answer. Again, the candidate uses 'carbon' when he means carbon-containing compounds or carbon dioxide. There are four marks available and there are not four ideas offered to attempt to get those marks. In longer questions, you must try to match your answer with the mark allocation. Do not simply write until the space is full: spot ideas that may be worth marks. He has also tried both sides of the same argument. This happens when you plough – this happens when you do not. There is a logical sequence of events resulting from reducing ploughing: there is a reduction of the amount of oxygen in the soil; this reduces the activity of decomposers; less aerobic respiration occurs; less of the organic material in the soil is broken down; less carbon dioxide is released. Think logically and do not miss steps.

(ii) Ploughing can increase the activity of nitrifying bacteria in the soil. Explain how ploughing can do this and how the activity of nitrifying bacteria can benefit crop plants. *(5 marks)*

Ploughing mixes the nitrifying bacteria with ammonium ions in the soil. These are converted to nitrites and nitrates. Nitrates can benefit crop plants as nitrogen can be absorbed by plants in this form. Plants take this up in their roots and take it to the parts that need it. The nitrogen is used to make protein and protein is used in all living things for growth and replacement. So with more nitrogen the plant makes more protein and grows more.

This answer is poorly expressed but two ideas would get credit; the absorption of nitrates by plants; the use of the nitrogen to make protein. However, several steps have been missed. Ploughing adds oxygen to the soil; this is vital for nitrifying bacteria as they are aerobic. You must also remember the steps in the nitrogen cycle. Ammonium ions are converted to nitrite ions and then to nitrate ions. If you want to give the formulae of the ions instead of their names, that is acceptable but not necessary. If you give both names and formulae, make sure you get the formulae right. If you don't, you may lose the mark for the name of the ion. Ammonium ions are NH_4^+, nitrite ions are NO_2^- and nitrate ions are NO_3^-. Be sure you make it clear that nitrites are produced and then nitrates. This answer suggests that they are both made from ammonium ions, which is not correct.

(AQA 2002)

EXAMINATION QUESTIONS

1 Read the following passage.

Early settlers used a technique known as 'slash and burn' to clear land for growing crops. Trees were cut down and burned and seeds of crop plants were scattered on the cleared land. After a few years, crop growth was usually so poor that people would move on and repeat the process. At low human population densities there was no long-term damage to the forest as the cleared areas of land had a chance to recover once people had left.

With an increase in human population, and over periods of time, large areas of forest have been destroyed. Modern developments have made possible greater yields from an area of land and farming has become more intensive. To maintain soil fertility, farmers now add fertilisers to the soil.

Use information from the passage and your own knowledge to answer the following questions.

a) Explain how the process of 'slash and burn' would affect the availability of carbon in the atmosphere. *(2 marks)*

b) Explain how bare, cleared land could once again become forest. *(3 marks)*

c) Fertiliser, such as manure, contains ammonium compounds. Explain how the presence of soil bacteria and the use of manure improve crop yield. *(6 marks)*

d) Explain the advantages of conserving a forest ecosystem. *(4 marks)*

(AQA 2003)

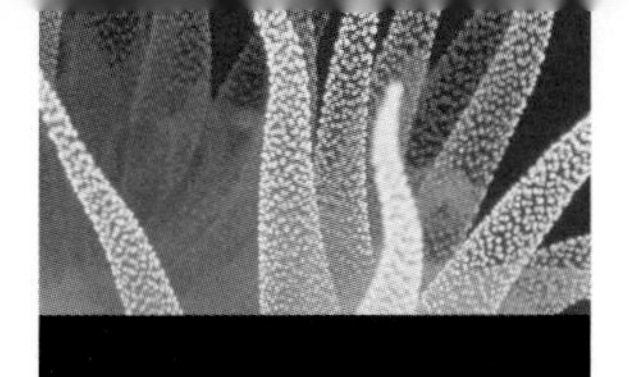

7 Uptake and loss of water in plants

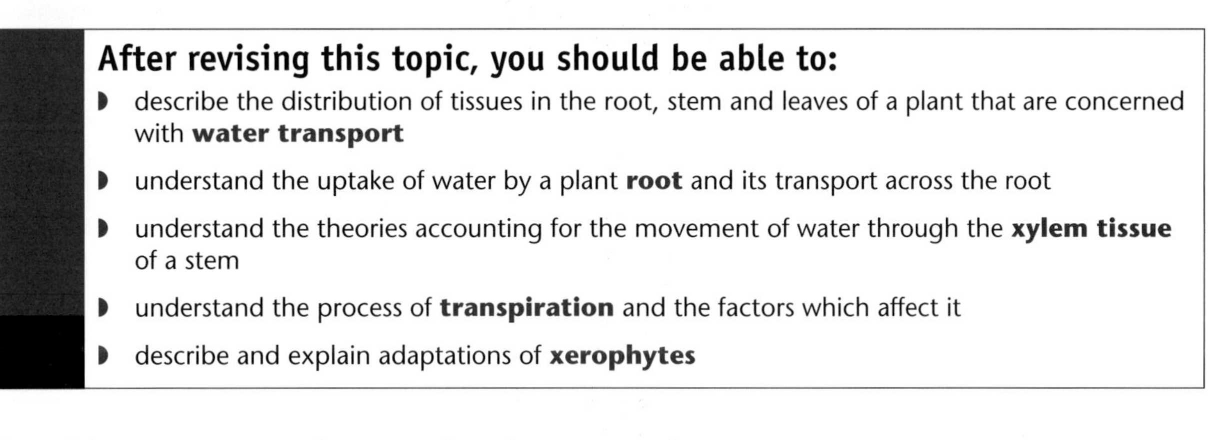

After revising this topic, you should be able to:

- describe the distribution of tissues in the root, stem and leaves of a plant that are concerned with **water transport**
- understand the uptake of water by a plant **root** and its transport across the root
- understand the theories accounting for the movement of water through the **xylem tissue** of a stem
- understand the process of **transpiration** and the factors which affect it
- describe and explain adaptations of **xerophytes**

Figure 7.1 gives an overview of water uptake and transport in plants.

Uptake of water by **epidermis** in younger parts of roots

↓

Transport of water from epidermis to xylem tissue in the **central vascular bundle** of root

↓

Water moved through xylem from root, up stem and to leaves

↓

Water crosses the tissues in the leaves and is lost through **stomata**

FIGURE 7.1 An overview of water uptake and transport in plants.

Structure of a root

Roots anchor plants in the ground. Only the **younger ends** of the root are permeable and this is where water is taken up by **osmosis**.

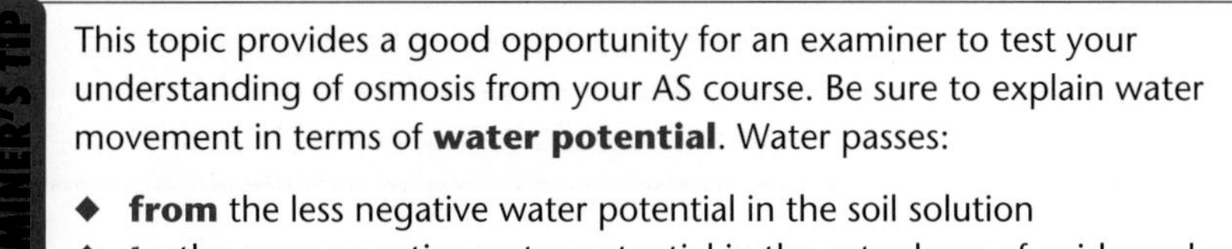

EXAMINER'S TIP

This topic provides a good opportunity for an examiner to test your understanding of osmosis from your AS course. Be sure to explain water movement in terms of **water potential**. Water passes:

- **from** the less negative water potential in the soil solution
- **to** the more negative water potential in the cytoplasm of epidermal cells surrounding the root
- **through** the permeable cell walls and the partially permeable membranes of the epidermal cells of the root.

Figure 7.2 represents the structure of a root. Note the following structures.

- **Epidermis** – a single layer of cells around the root
- **Root hairs** – extensions from some epidermal cells that **increase the surface area** for water uptake
- **Cortex** – the tissue that makes up the bulk of the root
- **Endodermis** – a **single layer** of cells surrounding the central vascular bundle. These cells have walls containing a band of **suberin** – a waxy substance that is impermeable to water – in the **Casparian strip**
- Central vascular bundle, containing **xylem** which transports water and **phloem** which transports organic substances

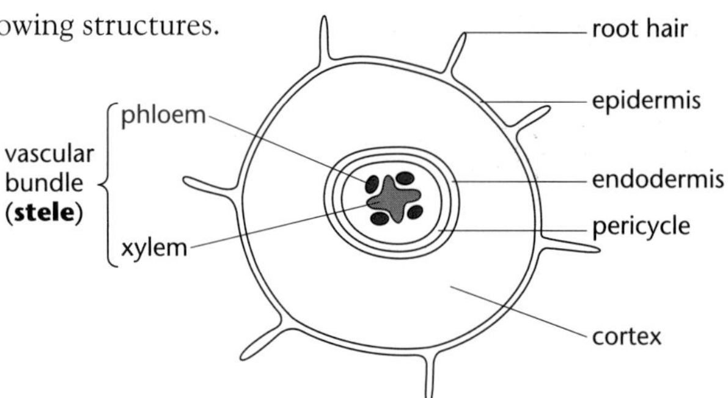

FIGURE 7.2 A transverse section through a young root.

Uptake of water and its transport across the root

Table 7.1 and Figure 7.3 summarise the uptake of water and its movement through a root.

Pathway of water	Process by which water is moved
Soil solution into epidermis	The water potential of epidermal cells is **lower** (more negative) than the soil solution around them. Water moves from the soil solution into the epidermal cells by **osmosis**.
Epidermis to cortex	**Apoplast pathway** – **cellulose** cell walls are permeable to water. Water moves through adjacent cell walls in the spaces between their fibres of cellulose. **Symplast pathway** – water moves by **osmosis** from the cytoplasm of one cell to the cytoplasm of an adjacent cell. **Gaps in the cell walls (plasmodesmata)** reduce the distance for osmosis between cells.
Through endodermis	The **Casparian strip** prevents water moving through the cell walls of the endodermis. All water moves into the **cytoplasm of the endodermal cells** (see Figure 7.4).
Endodermis to xylem	Cells of the endodermis use energy to secrete **ions** into the xylem. This **lowers the water potential** in the xylem so that water passes from the cytoplasm of the endodermal cell into the xylem by osmosis. This process causes a pressure (called **root pressure**) in the xylem.

TABLE 7.1 The processes by which water moves from the soil solution into the xylem of a root.

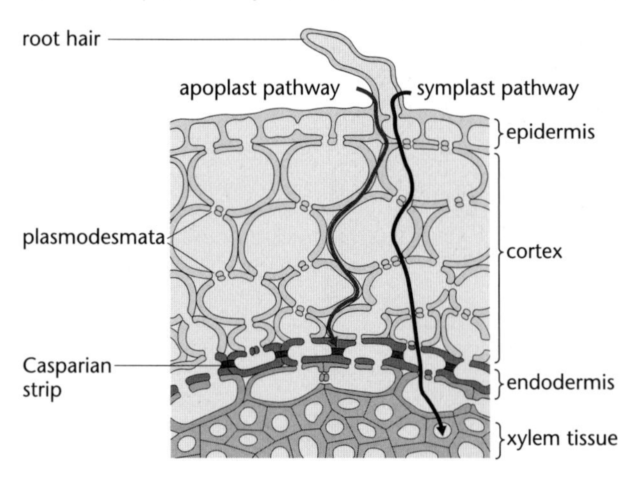

FIGURE 7.3 Water moves from the epidermis to the endodermis of a young root by two pathways: the apoplast pathway and the symplast pathway.

1 Explain why the cytoplasm of epidermal cells has a lower (more negative) water potential than the solution in the soil.

2 Use Figure 7.3 to explain the way in which plasmodesmata affect the rate of water movement across the cortex of a root.

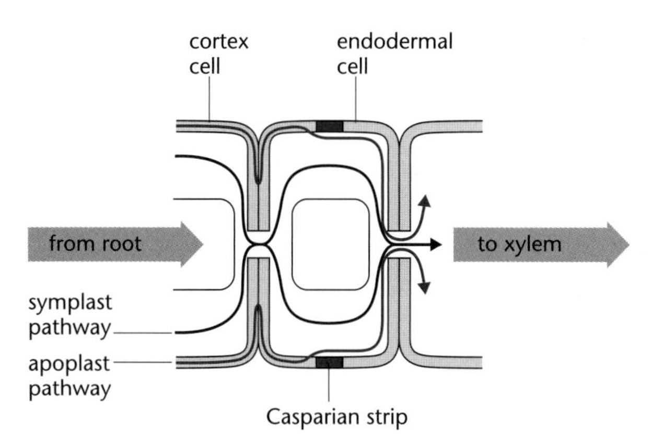

Figure 7.4 The Casparian strip stops water passing the endodermis via the apoplast pathway.

Movement of water from root to leaves

Water is carried to the leaves from **xylem tissue** in the **root** through **xylem tissue** in the stem.

Xylem tissue is like a series of continuous tubes. The tubes are called **vessels** and are made up of **vessel elements** arranged end-to-end. Figure 7.5 shows how these vessel elements:

- have cell walls that are impregnated with lignin. This strengthens the walls and makes them waterproof.
- have small pits in their cell walls. These are regions of cell wall without lignin. Since cellulose is permeable, water can move sideways from vessel to vessel through these pits.
- contain no cytoplasm. Instead, each wall surrounds a **lumen** which contains a solution of water and mineral ions. The solution forms a **continuous column** up the length of each vessel.

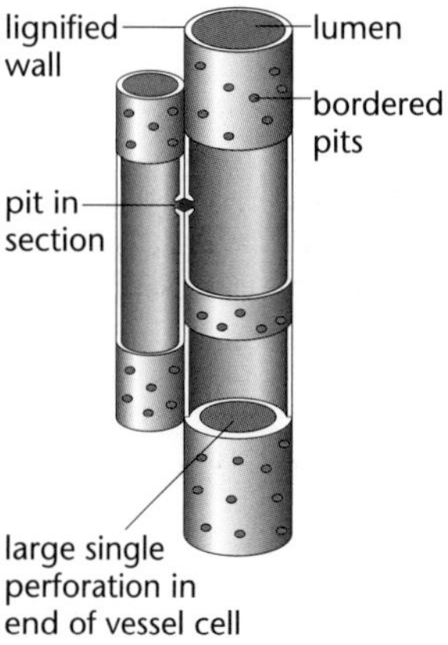

lignified wall
lumen
bordered pits
pit in section
large single perforation in end of vessel cell

Figure 7.5 Adjacent xylem vessels. These vessels form a tube inside the stem of a plant within which is a continuous column of water-based solution.

Xylem tissue is found in **vascular bundles** in the stem. Figures 7.6 shows how these bundles:

- contain **xylem** and **phloem** (which transports mainly organic substances)
- are arranged in a **ring** around the stem.

?

3 **Explain why the solution in the xylem contains mineral ions.**

4 **Suggest one mechanical advantage conferred by the arrangement of vascular bundles in the stem.**

5 **Name the force that: a) enables water molecules to be pulled up the xylem in the stem; b) occurs in the xylem during transpiration**

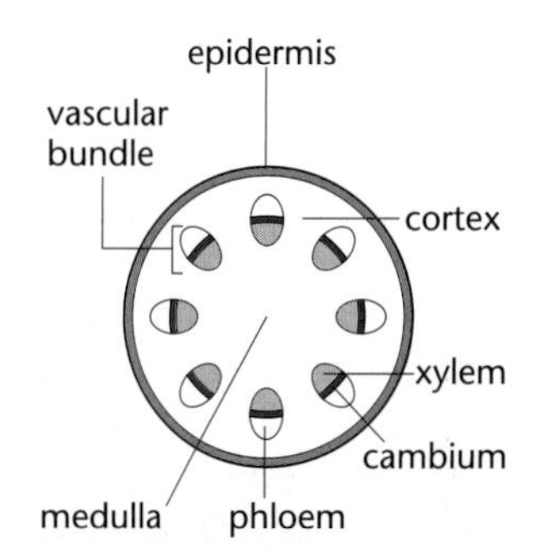

Figure 7.6 A transverse section through a herbaceous stem. Xylem is found in vascular bundles which form a ring around the stem.

Theories to explain movement of water up stem

Table 7.2 summarises three theories that could account for the movement of water through the xylem of a stem.

Name of process	Explanation	Evidence	Notes
Root pressure	◆ Because of Casparian strip, water in root must pass through cytoplasm of endodermal cells. ◆ Active transport of ions by endodermal cells into xylem creates a **water potential gradient** that pushes water from root into, and up, xylem in stem.	◆ Sap exudes from xylem of a stem that is cut close to the ground. ◆ Metabolic inhibitors, low temperatures and lack of oxygen in the root all reduce root pressure.	◆ Maximum measured values of root pressure are about 150 kPa. ◆ This is not enough to push water to the top of tall trees.
Capillarity	If a fine tube is dipped in water, the water rises up the tube by **capillarity** as water molecules adhere to the **sides** of the tubes.	Xylem vessels are like fine capillary tubes (about 20 μm diameter). Water rises by capillarity in glass tubes of this diameter.	The rise of water by capillarity is less than 50 mm and so is not strong enough to transport water alone.
Cohesion-tension	◆ Put simply, **loss of water from the leaves pulls more water up the xylem in the stem.** ◆ Water is lost from xylem vessels in the leaves during transpiration. ◆ Neighbouring water molecules form hydrogen bonds, making them stick together. This is called **cohesion.** ◆ Water molecules also **adhere** to the sides of the xylem vessels. ◆ As water is lost from xylem in the leaves, cohesion causes more water to be pulled up the xylem vessels. This causes **tension** (an inward pressure) in the xylem.	◆ Tension has been measured in the xylem as plants transpire. ◆ Lignified walls of narrow xylem vessels are strong enough to withstand the tension measured. ◆ The diameter of trees reduces when they are transpiring (tension pulls xylem vessels inwards) and increases when they are not transpiring (reduced tension enables xylem vessels to 'relax'). ◆ Air bubbles in water column within xylem interfere with cohesion between water molecules and stop upward movement of water.	◆ The measured pressures are strong enough to account for the movement of water to the top of the tallest trees.

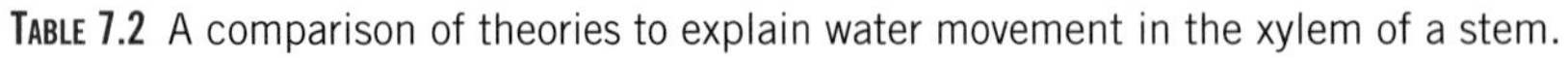

TABLE 7.2 A comparison of theories to explain water movement in the xylem of a stem.

EXAMINER TIP

An examiner could ask you a recall question (AO1) about the theories accounting for water movement up the xylem of a stem. However, you could be asked to interpret data from investigations into water movement. To do this successfully, you must understand the concepts that are involved.
During your revision, don't rely solely on memorising facts and concepts (AO1). Ensure that you test your understanding and practise data interpretation as well (AO2). These skills are tested more in the A2 examinations than they were in the AS examinations.

? 6 If cut flowers are left in the air for a while, they often wilt even when placed in a vase of water. Suggest why.

Movement of water through leaf

Figure 7.7 shows the internal structure of a leaf. Because **water vapour** is lost by **diffusion** through **open stomata**, there is a water potential gradient through the whole leaf.

- Water vapour diffuses from the air spaces in the leaf into the surrounding air through open stomata. This is called **transpiration**.
- Water lost from the air spaces in the leaf is replaced by **evaporation** of water from the cell walls of the **mesophyll cells** surrounding the air spaces.
- As their water potential becomes more negative, mesophyll cells gain water by **osmosis** from adjacent cells and, ultimately, from xylem vessels in the leaf.

? 7 Why will the water potential of mesophyll cells become more negative?

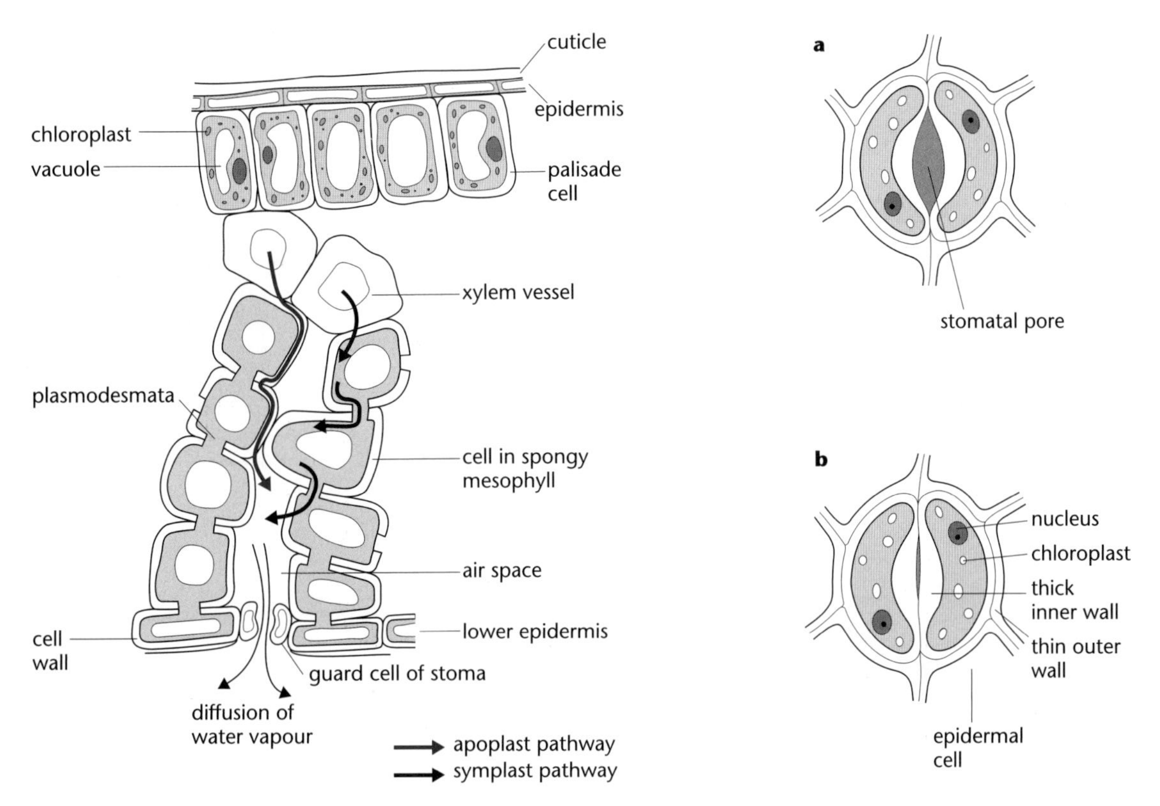

FIGURE 7.7 A vertical section through a leaf showing the pathway of water.

FIGURE 7.8 The stomatal pore is a) opened and b) closed by guard cells.

The diameter of each stomatal pore is controlled by two **guard cells**. Unlike the other epidermal cells, these cells have chloroplasts, are not linked to other cells by plasmodesmata and have unevenly thickened walls (Figure 7.8).

- When the guard cells take in water and become turgid, they become more curved and **increase** the diameter of the stoma between them.
- When the guard cells lose water, they become less curved and **decrease** the diameter of the stoma between them.

Figure 7.9 shows variation in the rate of transpiration throughout a twenty four hour period.

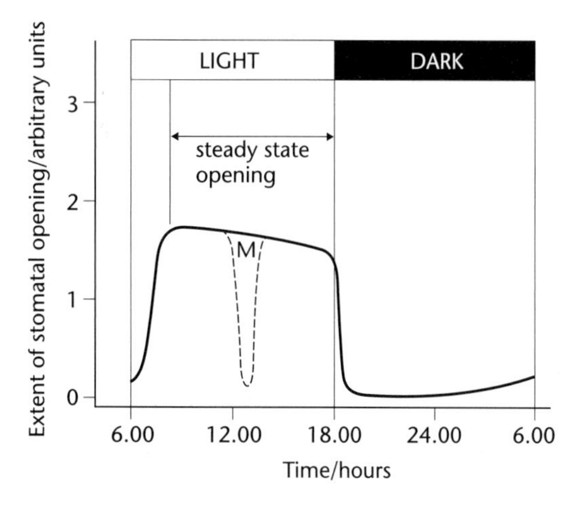

FIGURE 7.9 The rate of transpiration varies throughout the day.

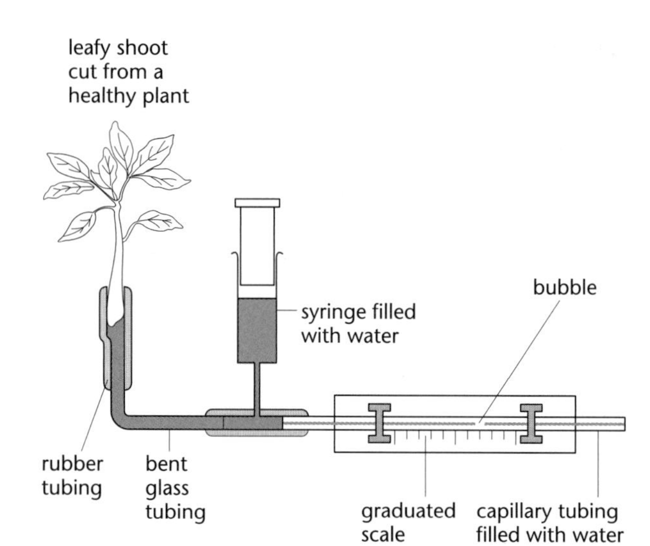

FIGURE 7.10 A simple potometer can be used to measure the rate of transpiration.

MEASUREMENT OF TRANSPIRATION

Figure 7.10 shows a **potometer** that can be used to measure the rate of transpiration.

- Water lost from the shoot's leaves during transpiration is replaced by water in the xylem.
- Provided cutting the stem did not introduce an air bubble in the xylem, the cut shoot takes up water from the potometer in **direct proportion** to the amount that it loses from its leaves during transpiration.
- The capillary tube and scale enable us to measure the uptake of water from the potometer.

> **? 8 Suggest why a leafy shoot is cut under water before being placed into a potometer.**

FACTORS AFFECTING TRANSPIRATION

Table 7.3 summarises the major environmental factors affecting transpiration rates.

Factor	Explanation
Humidity	If air around the leaf is **saturated** with water vapour, the diffusion gradient of water from leaf to air is less steep. As a result, the rate of diffusion decreases.
Temperature	◆ High temperatures increase the rate of **random thermal movement** of water molecules and so increase their rate of diffusion. ◆ In temperate countries, air of high temperature is generally **drier** than cool air. Dry air around the leaf increases the diffusion gradient of water from the leaf.
Wind speed	**Air currents** remove air from the leaf, increasing the diffusion gradient of water from the leaf.
Light intensity	◆ Guard cells tend to become turgid during daylight. As a result, the **stomata are open during daylight hours**. ◆ The mechanism of stomatal opening and closure is complex. It is most probable that light (particularly blue light) activates proton pumps in the guard cells. Pumping ions into their cytoplasm causes guard cells to take up water by osmosis and swell.

TABLE 7.3 Environmental factors affecting transpiration rate.

Xerophytes

Most water is lost from plants via the stomata; some is lost by evaporation from the surface of the plant.

Xerophytes are plants that live in dry regions. Table 7.4 summarises some of the adaptations of xerophytes that enable them to inhabit dry areas, including deserts and areas where the soil is frozen. Although most water is lost from plants via their stomata, some is lost through the **general surface area of the leaf**. Consequently, **most adaptations concern leaves**.

Feature	Explanation	Examples
Thick cuticles on leaves	Wax waterproofs the epidermis and reduces evaporation	Holly, rubber plant
Leaves reduced to small spines	Small leaves have few stomata and so the rate of transpiration is low	Conifers, cacti
Leaves curled to form a tube with the lower leaf surface on the inside of the tube	Air inside the tube becomes humid, reducing the rate of transpiration	Marram grass
Stomata are sunk below the epidermis of the leaves	The additional air space around the stomata becomes saturated with water, reducing the diffusion gradient out of the leaf	Marram grass
Stems are swollen with water-storage tissue	Specially adapted tissue stores water	Cacti

TABLE 7.4 Some of the adaptations of xerophytic plants.

WORKED EXAM QUESTIONS

1 a) Root pressure is a force that is partly responsible for the movement of water through xylem in stems. Explain how active transport of mineral ions into xylem vessels in the root results in water entering these vessels and then being moved up the xylem tissue. *(5 marks)*

Water enters the plant root through its root hairs. These increase the area of the root for more water uptake. Once in the root hair cells, the water travels across the root by the apoplast pathway and the symplast pathway. In the apoplast pathway, water travels through the spaces in the cell walls. In the symplast pathway, water travels through the cytoplasm of cells in the root. Eventually, the water gets to a layer of cells called the endodermis. The cell walls of these cells are waterproof (Casparian strip) and stop the apoplast pathway. All the water must travel into the cytoplasm of the endodermal cells. It is these cells which secrete mineral ions into the xylem vessels and cause water to move up the xylem by root pressure.

The candidate has made a common mistake – he has learned a topic in advance and written his learned response without reference to the question. He has given a reasonable account of water uptake and its movement across a root, although he has not mentioned water potential and osmosis, but has not answered the question. The question starts with the secretion of ions into the xylem – the candidate's answer ended here.

The mark scheme was: secretion of mineral ions into xylem reduces water potential in xylem; surface membranes of surrounding cells are partially permeable; water enters xylem by osmosis; down a water potential gradient; volume of water in xylem increases; pressure in xylem increases (and forces water upwards); water cannot move back out of xylem because of water potential gradient.

b) The presence of an air bubble in a xylem vessel in the stem blocks movement of water through that vessel. Use the cohesion-tension theory to explain why. *(4 marks)*

Because water molecules bond together with hydrogen bonds, there is a continuous column of water in the xylem vessels of the stem. This is called cohesion of water molecules. When water is lost from the leaves during transpiration it causes a tension (negative pressure) so that the column of water is pulled up the xylem to the leaves. If there is an air bubble in a xylem vessel it breaks the cohesion between water molecules in the column so that the water can no longer be pulled up in the transpiration stream.

This is a good answer and gains all four marks. Four marks maximum were awarded for: cohesion between water molecules; column of water in xylem; transpiration/evaporation of water from leaves; causes negative pressure/tension/pulls water up; air bubble prevents cohesion/breaks column.

Since the question asked candidates to 'use the cohesion-tension theory to explain why', no marks were awarded merely for mentioning the cohesion-tension theory.

c) Water vapour diffuses through open stomata into the atmosphere. Describe **two** structural adaptations of the leaves of xerophytes that reduce this loss. Using

Fick's law, explain how these two adaptations reduce the rate of diffusion of water vapour into the atmosphere. (6 marks)

Fick's law states that the rate of diffusion is proportional to

$$\frac{\text{area of exchange surface} \times \text{difference in concentration}}{\text{thickness of exchange surface}}$$

Plants can reduce their water loss by decreasing the area through which they lose water, decreasing the water vapour concentration gradient out of their leaves and increasing the thickness of the exchange surface.

Cactuses have a small number of spiny leaves. This gives their leaves a smaller surface area and reduces the rate at which they lose water. Rubber plants have a very thick layer of wax on the epidermis of their leaves and this increases the thickness of the exchange surface over which diffusion takes place. As Fick's law shows, this makes diffusion slower.

Again, a good answer which gains a full six marks. The candidate gained a mark for correctly stating Fick's law and a second mark for relating features of the equation to adaptations which reduce the rate of water loss. He gained two marks for each of the two adaptations he used – one mark for its description and one for stating the effect of this adaptation.

(AQA 2003)

EXAMINATION QUESTIONS

1 The diagram shows the pathway taken by water passing through a plant.

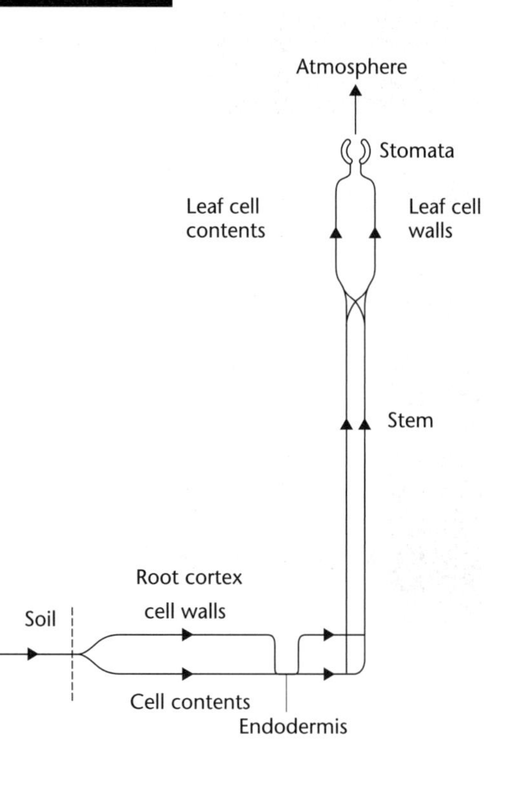

a) Name

(i) the process by which water enters root hairs from the soil. *(1 mark)*

(ii) the pathway through the cell walls of the root cortex. *(1 mark)*

b) All water passes through the endodermis by the same pathway. Explain what causes this. *(2 marks)*

c) Describe and explain how water moves through the trunk of a tree to the leaves. *(6 marks)*

d) The graph shows the rate of water flow through a branch near the top of a tree on a summer's day.

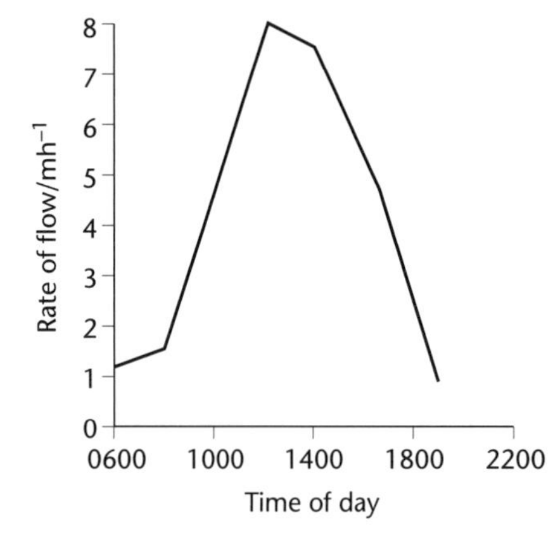

Describe and suggest explanations for the changes in the rate of flow during the day. *(4 marks)*

(AQA 2003)

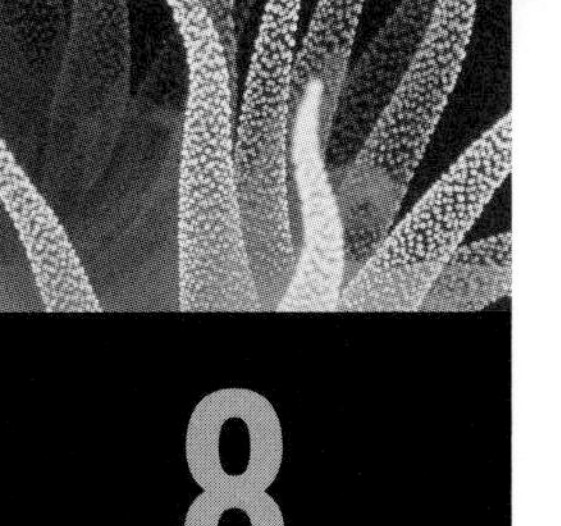

8 Liver and kidney

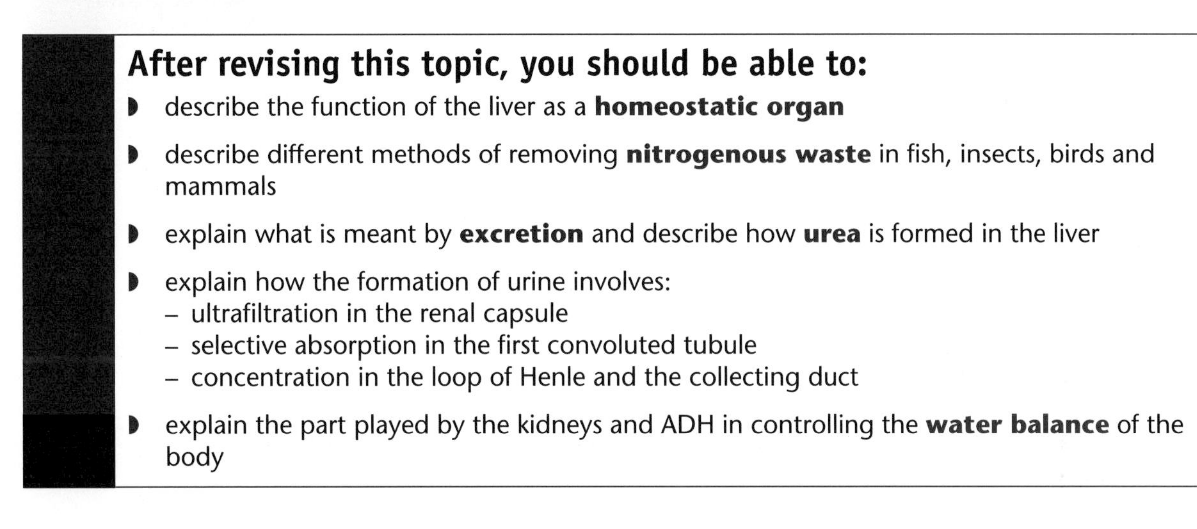

After revising this topic, you should be able to:

- describe the function of the liver as a **homeostatic organ**
- describe different methods of removing **nitrogenous waste** in fish, insects, birds and mammals
- explain what is meant by **excretion** and describe how **urea** is formed in the liver
- explain how the formation of urine involves:
  - ultrafiltration in the renal capsule
  - selective absorption in the first convoluted tubule
  - concentration in the loop of Henle and the collecting duct
- explain the part played by the kidneys and ADH in controlling the **water balance** of the body

The liver

STRUCTURE

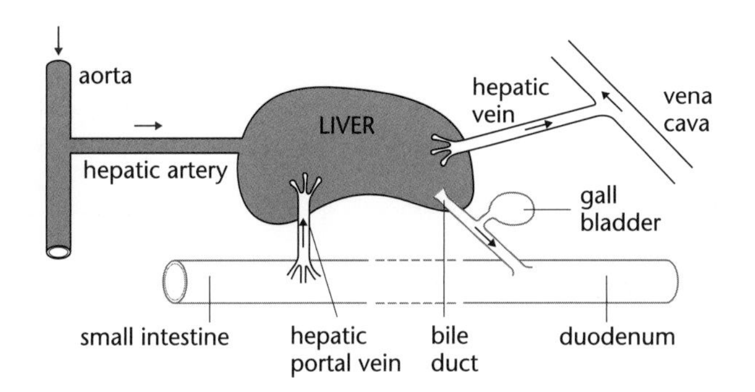

FIGURE 8.1 Structure of the liver.

- The liver is richly supplied with blood from the:
  - HEPATIC ARTERY – this delivers **oxygenated blood** to the liver cells
  - HEPATIC PORTAL VEIN – this brings blood to the liver **from the small intestines**, so it is rich in glucose and amino acids (after they have been absorbed following digestion)
- Blood is **removed** from the liver by the HEPATIC VEIN.
- A further vessel associated with the liver is the BILE DUCT – this takes the **bile** from the liver, where it is produced, to the **gall bladder** where it is stored. Bile is released into the small intestine and emulsifies triglycerides.

EXAMINER'S TIP

You will not have to remember the names of many blood vessels in the body but all blood vessels associated with the liver start with 'hepatic' and those associated with the kidney start with 'renal'.

HOMEOSTATIC FUNCTION OF THE LIVER

The liver has many functions including the production of bile, the breakdown of old red blood cells and the storage of vitamins and iron. However, there are only three functions which are homeostatic – **keeping conditions in the body constant**.

Homeostatic function	Details
1. Carbohydrate metabolism	◆ Glucose converted to glycogen reduces glucose concentration in the blood. ◆ Glycogen converted to glucose increases the concentration of glucose in the blood. ◆ Non-carbohydrates (such as amino acids and glycerol) converted to glucose increases glucose concentration in the blood.
2. Lipid metabolism	◆ Processes fatty acids which can be transported and deposited in the body. ◆ Makes or removes cholesterol and phospholipids from the blood. ◆ Excess cholesterol and phospholipids are removed in the bile.
3. Amino acid metabolism	◆ Excess protein/amino acids cannot be stored. ◆ The liver breaks down the excess by deamination.

1 What is the difference between essential and non-essential amino acids?

Nitrogenous excretory products

Excretion is the **removal of waste products** resulting from metabolic processes in an organism. They include carbon dioxide, a waste product of respiration.

Other waste products are:

- ammonia
- urea
- uric acid

These all contain nitrogen, so are called **nitrogenous** waste products.

2 Name another chemical that is a waste product of respiration.

Nitrogenous waste product	Notes
Ammonia	◆ It is extremely soluble in water ◆ It is very toxic and cannot be stored ◆ It can be removed if highly diluted so lots of water is lost in getting rid of it ◆ Freshwater fish excrete ammonia
Urea	◆ It is less soluble in water than ammonia ◆ It is less toxic than ammonia ◆ Less water is lost getting rid of urea ◆ Mammals excrete urea
Uric acid	◆ It is extremely insoluble in water compared to urea ◆ It is much less toxic than urea ◆ Very little water is required for its excretion ◆ Birds and reptiles excrete uric acid

3 Give one disadvantage and one advantage to a terrestrial animal of excreting ammonia.

The kidneys

STRUCTURE

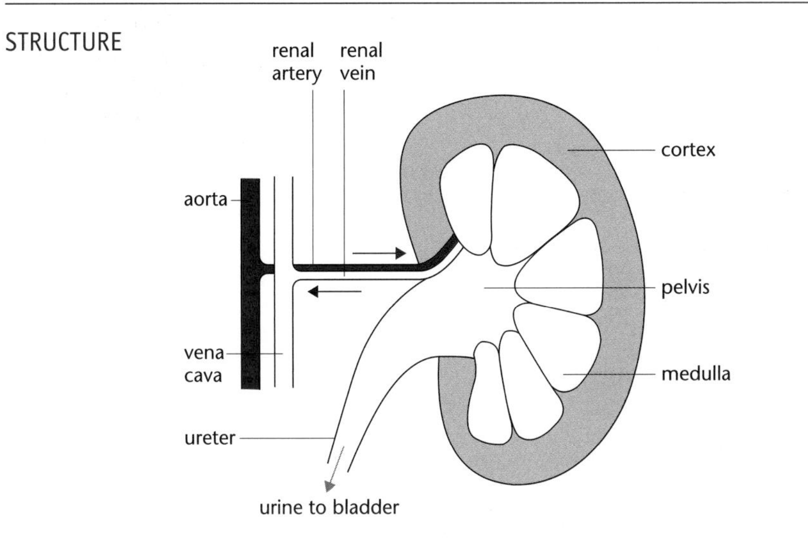

Figure 8.2 Structure of the kidney and its vessels.

- The kidneys receive blood from the **renal arteries** – these supply the kidney cells with oxygen and nutrients.
- Blood is removed from the kidneys by the **renal vein.**
- The **ureter** removes urine from the kidney, taking it to the **bladder** where the urine is stored before it is expelled.

URINE FORMATION

- Urine is formed in the kidney in millions of kidney **tubules**
- Each kidney tubule (**nephron**) is divided into distinct regions, each having a particular function.
  - **Renal capsule** – for ultrafiltration
  - **First convoluted tubule** – for reabsorption
  - **Loop of Henle, second convoluted tubule** and **collecting duct** – all play important roles in the reabsorption of water

4 What is the difference between urea and urine?

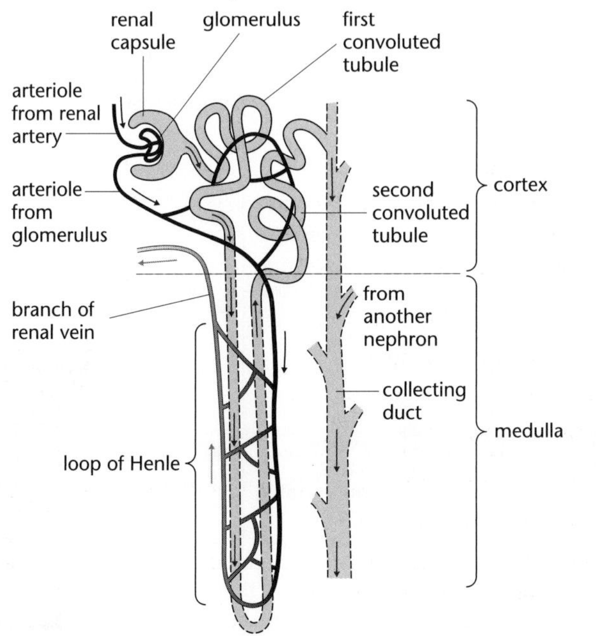

Figure 8.3 Structure of a kidney tubule.

Function	Region	Details
Ultrafiltration	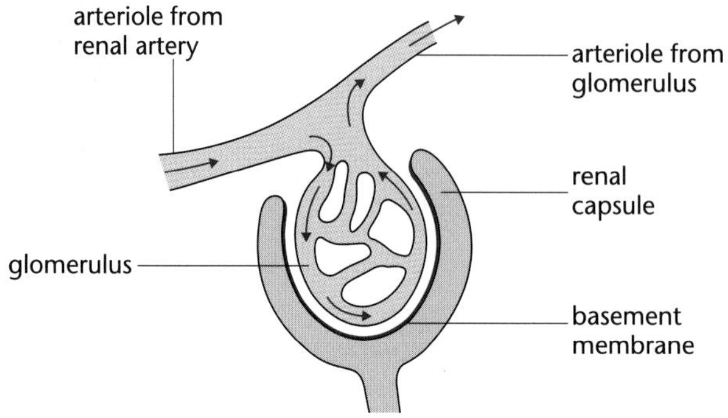 arteriole from renal artery arteriole from glomerulus renal capsule glomerulus basement membrane	◆ The pressure of blood in the glomerulus forces fluid out of the capillary across the **basement membrane** and into the renal capsule ◆ The basement membrane acts as the filter ◆ Only molecules small enough can cross the basement membrane. These include glucose, amino acids, salts, water and urea ◆ Protein molecules are too large to pass through basement membrane
Reabsorption	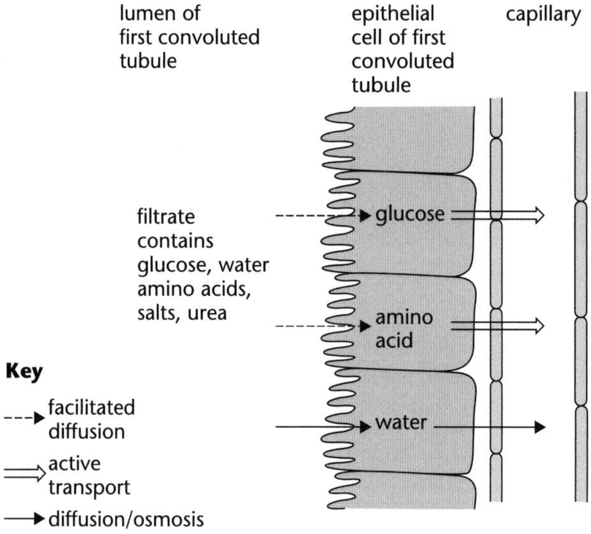 lumen of first convoluted tubule epithelial cell of first convoluted tubule capillary filtrate contains glucose, water amino acids, salts, urea glucose amino acid water **Key** facilitated diffusion active transport diffusion/osmosis	◆ Useful substances such as glucose and amino acid are immediately **reabsorbed** into the blood. This is by a combination of facilitated diffusion and active transport ◆ Water is reabsorbed by osmosis ◆ Urea is not reabsorbed; but because water has been removed, its concentration in the tubule increases
Reabsorption of water	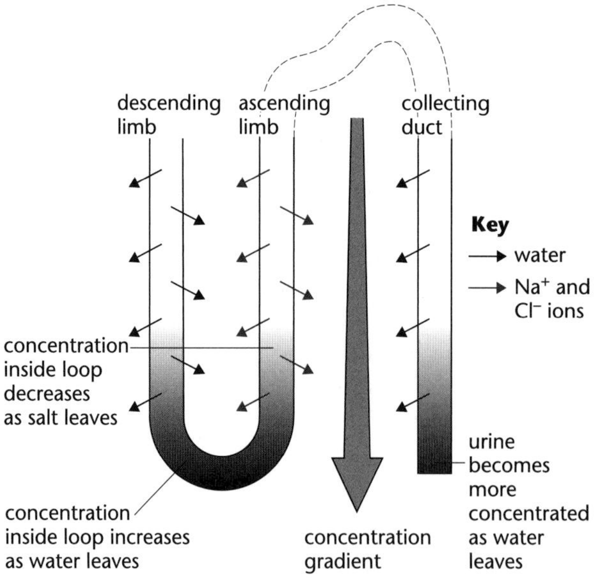 descending limb ascending limb collecting duct **Key** water Na^+ and Cl^- ions concentration inside loop decreases as salt leaves urine becomes more concentrated as water leaves concentration inside loop increases as water leaves concentration gradient	◆ The **descending** loop of Henle loses water by osmosis: the filtrate concentration increases ◆ The **ascending** loop of Henle loses salts, sodium ions and chloride ions, (passively and actively) into the medulla: the filtrate concentration decreases ◆ Movement of ions out of the loop of Henle results in a concentration gradient in the medulla ◆ When the filtrate travels down the **collecting duct**, because the filtrate always has a lower water potential than the cells in the medulla, water continually moves out by osmosis ◆ The result is a concentrated urine – lots of urea in the smallest amount of water

Remember: The **longer** the loop of Henle, the **greater** the concentration gradient that can be established in the medulla resulting in more water being reabsorbed and a more concentrated urine being produced.

The kidneys and water balance

In different conditions, different amounts of water are lost.

- In hot conditions where large amounts of water are lost in sweat, it is important that as little as possible is lost as urine.
  - The **loss of water** lowers **the water potential** of the blood.
  - This is detected by **osmoreceptors** in the **hypothalamus**.
- In conditions where the body has gained too much water, it is important that the body loses this excess in the urine.
  - The **excess water increases the water potential** of the blood.
  - This is detected by osmoreceptors in the hypothalamus.

Anti-diuretic hormone (ADH) plays an important part in controlling the amount of water lost as urine.

- ADH is made in the **hypothalamus**.
- It is sent to the **pituitary gland**, where it is released into the blood.
- The second convoluted tubule and collecting duct have **ADH receptors** in their membranes.
- The presence of ADH increases the permeability of the second convoluted tubule and the collecting duct to water.

EXAMINER'S TIP

The kidney nephron and the functions of each part is hard to visualise. Do not be afraid to draw a simple diagram so that you can remember which part does what.

? **5 Explain why the cells of the first convoluted tubule**
a) contain many mitochondria
b) have microvilli.

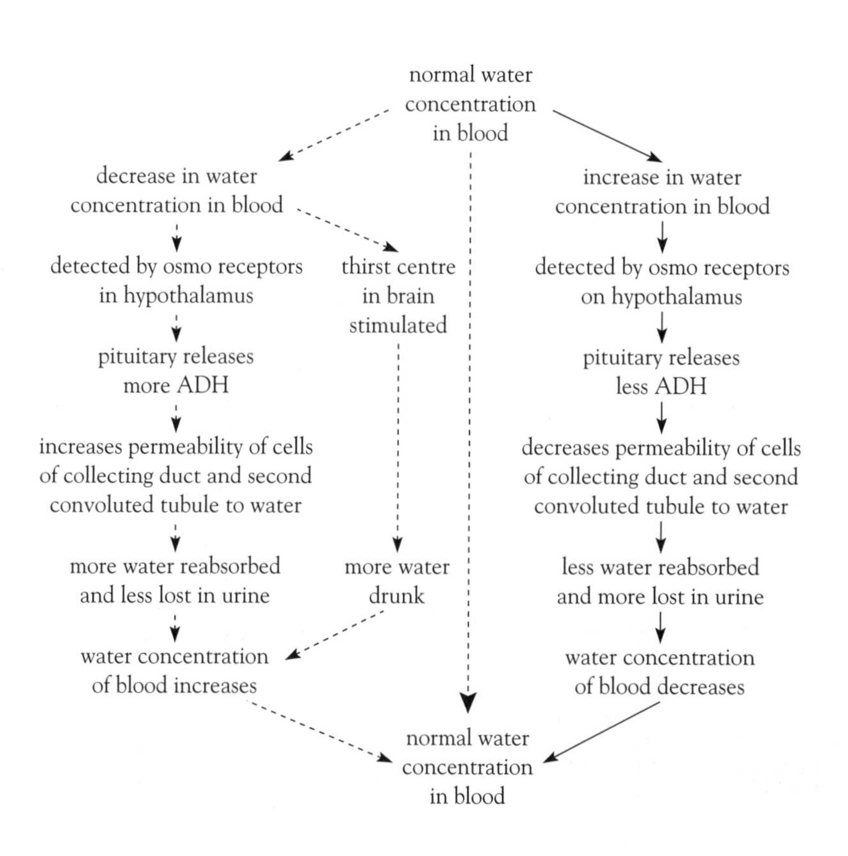

FIGURE 8.4 Summary of the effect of ADH.

WORKED EXAM QUESTIONS

1 Most nitrogenous waste material comes from surplus protein in the diet.

a) **Figure 1** shows some of the important steps in the formation of urea in mammals.

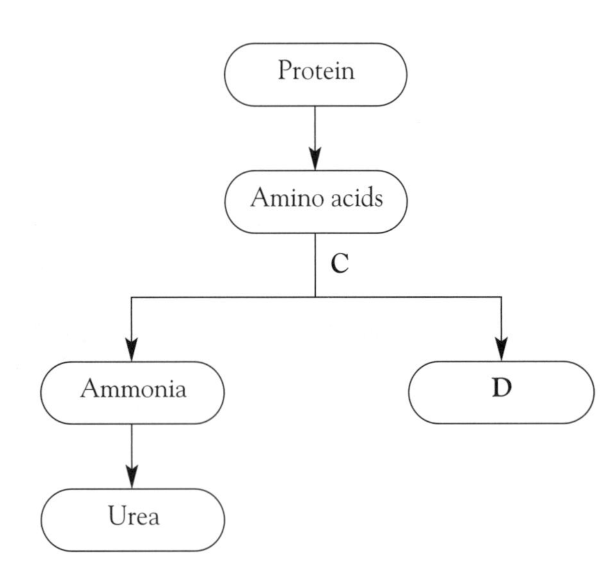

FIGURE 1

> There are a number of sections of this question highlighted by the letter **S**. This indicates that you will need to use material from other areas of the specifications and probably from your AS modules. Here knowledge of basic biochemistry and movement into and out of cells, from module 1, is necessary.

S (i) Why are protein molecules considered to be polymers? *(1 mark)*

A polymer is made up of many monomers. The monomers in protein are amino acids.

> The candidate obviously knows what a polymer is and how that definition can be applied to protein. A very good answer.

S (ii) Name process **C**. *(1 mark)*

The process C on the diagram is deamination.

> Again this is the correct answer, but if you are asked to 'name' a process or structure or molecule, there is no need to make a sentence of the answer. The word 'deamination' alone is perfectly acceptable.

(iii) Describe what happens to the part of the amino acid molecule labelled **D**. *(2 marks)*

This part of the amino acid is used for respiration. It is passed into the mitochondria where it is broken down releasing carbon dioxide and water. As this is aerobic respiration 38 molecules of ATP are produced which are used as an energy source.

The molecule itself cannot be used in respiration: it is first converted into either pyruvate or one of the intermediate compounds of the Krebs cycle. Those molecules, however, can be used in respiration. The candidate has tried too hard to give further information in an attempt to pick up his two marks. 38 molecules of ATP are produced from the aerobic respiration of one molecule of glucose – not from molecule D.

b) Tadpoles of the common frog live in freshwater ponds. Over a period of weeks, they undergo metamorphosis as they develop into adult frogs and move onto land. **Figure 2** shows the proportions of nitrogenous waste excreted as ammonia and as urea during the time after the tadpole hatches from the egg.

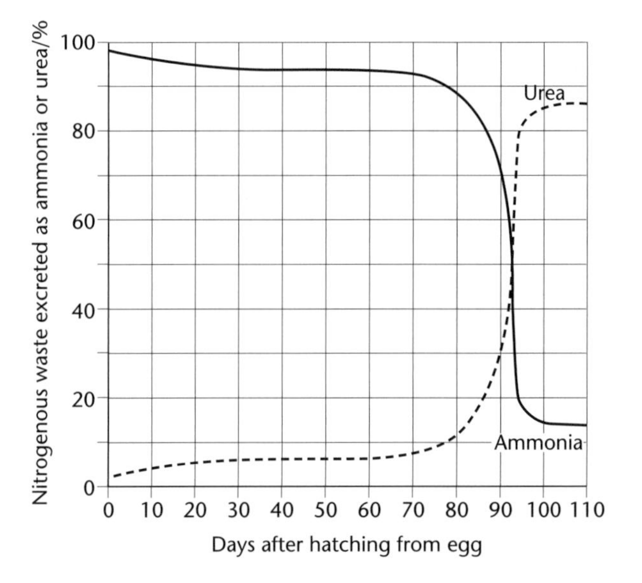

Figure 2

(i) Describe the changes in proportions of the two excretory products over the period shown. *(2 marks)*

The two excretory products are ammonia and urea. The amount of ammonia falls at the same time as the amount of urea rises.

Although he has described the change in the proportion of urea and ammonia, and would be awarded the first mark, he has not used the information on the graph. The changeover point is between day 60/70 to day 100. Always quote figures when you are given a graph or table of data. This was expected for the second mark.

(ii) Suggest an explanation for the changes in the proportion of ammonia and urea excreted. *(4 marks)*

After hatching the tadpole lives in water for a while and therefore can use ammonia as an excretory product. Ammonia is very poisonous but is safe when dissolved in a lot of water. If the tadpole lives in water it can pass the ammonia out safely. The adult frog moves onto land and ammonia is therefore going to be hard to get rid of without wasting a lot of water. To conserve water the frog will use urea as an excretory product instead which does not need as much water to get rid of.

It is clear that the candidate knows that the principle here is the movement of the animal from an aquatic to a terrestrial habitat. The need to use a different excretory product to conserve water and the toxic nature of ammonia is also given. However, the fact that the excretory product must be stored by a land-living animal (but can be continuously excreted by an aquatic one) was not considered. He has also not used the information given in Figure 1: that ammonia is converted into urea.

S c) In the first convoluted tubule of a human nephron, sodium ions, glucose molecules and water molecules are reabsorbed into the blood plasma. **Figure 3** illustrates how these substances are reabsorbed.

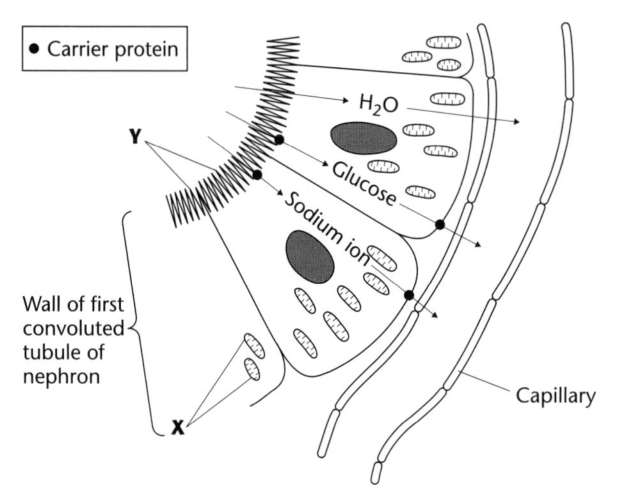

FIGURE 3

(i) Explain the large number of the organelles labelled **X** in these cells. *(2 marks)*

Organelle X is a red blood cell. These cells are necessary to provide the oxygen for aerobic respiration. Energy is needed for the processes in the tubule.

There are half-truths here, but the answer is not worth any marks. The organelles are mitochondria. Red blood cells would not be found outside the blood system; and certainly not inside other cells. He has not looked carefully at the labels on the diagram that make it clear that these organelles are in the cells of the convoluted tubule. It is important that you recognise the major organelles and, although the diagram does not show the two membranes, the typical cigar shape and internal projections (*cristae*) are clear enough indicators. The mitochondria are the site of aerobic respiration, and a second mark would be given for the reason 'the tubule cells need energy'. He does not link the presence of mitochondria and carrier proteins to their function of actively transporting glucose from the tubules into the capillary which require ATP.

(ii) Give **two** differences between the process by which water enters the capillary from the epithelial cell and that by which glucose and sodium leave the epithelial cell. *(2 marks)*

1 ***Water enters by osmosis but sodium and glucose do not.***

2 ***Sodium and glucose leave the cell by using carrier proteins but water does not.***

This question is looking for the difference between simple diffusion and active transport. So osmosis takes place passively; down a concentration gradient; without the help of carriers. This candidate certainly has the last point. Remember that when asked for a comparison, it is important to make clear to which molecule you are referring.

(iii) Explain the importance of the structures labelled **Y** on the epithelial cells. *(1 mark)*

Y are villi which increase the surface area.

Although he was not expected to name the structures Y, he has identified them incorrectly. Y are microvilli – projections of the cell membrane – and not villi, which are multicellular folds of the small intestine. Both do increase the surface area, but the question asks for an explanation. So why is an increased surface area an advantage to this cell? The full answer is: 'to increase the surface area for absorption'.

(AQA 2003)

EXAMINATION QUESTIONS

1 a) The flow chart summarises some of the events involved in the control of the water potential of blood plasma.

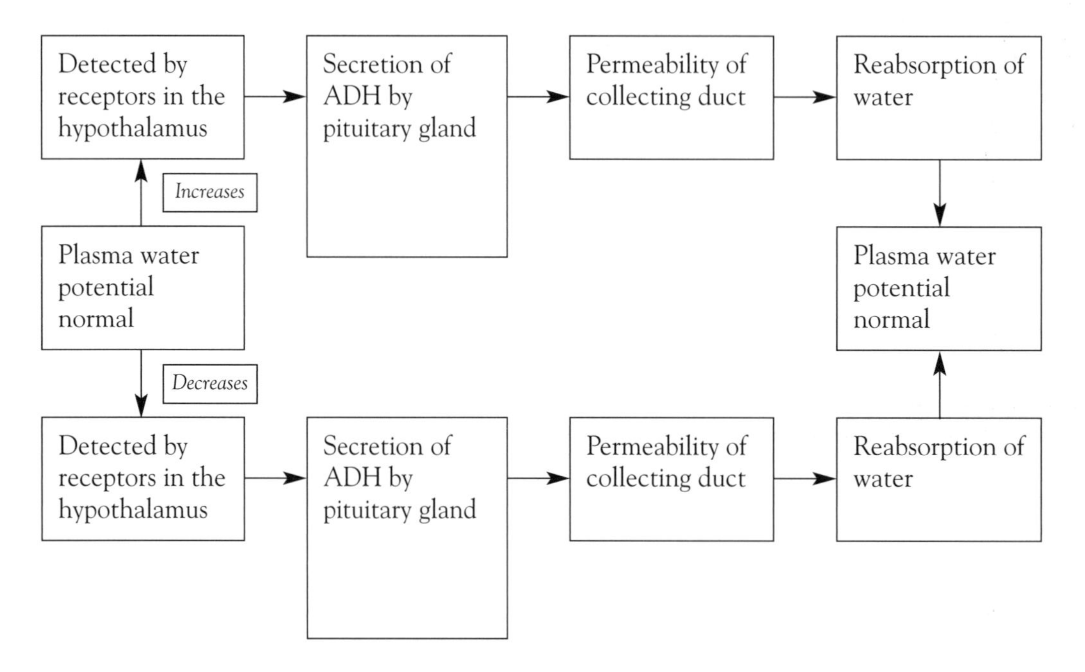

(i) Write 'increases' in **one** appropriate box and 'decreases' in **one other** appropriate box. *(1 mark)*

(ii) Give evidence from the flowchart which shows that the control of plasma water potential involves negative feedback. *(1 mark)*

b) Describe the role of the loop of Henle in the reabsorption of water from the collecting ducts. You may draw a diagram if it helps your answer. *(3 marks)*

(AQA 2003)

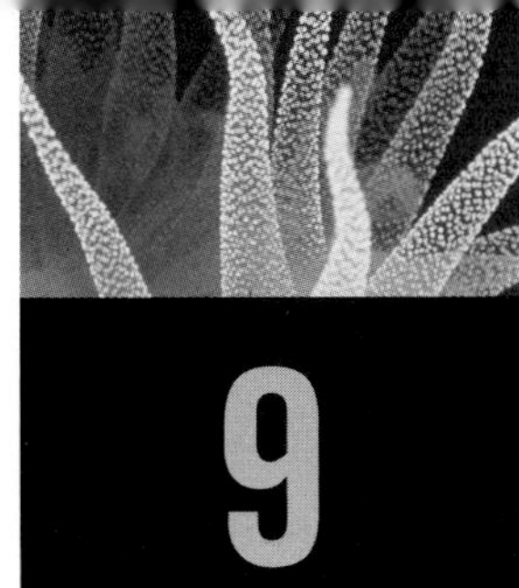

9 Homeostasis

After revising this topic, you should be able to:

- understand what is meant by **homeostasis** and explain why it is important in living organisms
- understand what is meant by **negative feedback** and explain how it is involved in homeostasis
- explain how blood glucose concentration is controlled in the body
- explain how temperature is controlled in ectotherms and endotherms

Why is homeostasis important?

Homeostasis is the mechanism that maintains a **constant internal environment**.

- It involves mechanisms, which keep conditions inside an organism within narrow limits and, thus, allow that organism to be independent from fluctuating external conditions.
- It is important that the cells of the body are bathed constantly with fluid at a **constant temperature, pH** and **water potential**, despite changes which tend to alter them.
  - Enzyme reactions are affected by changes in pH and temperature: extreme conditions could denature an enzyme.
  - Water moves in and out of the cell by osmosis. Constant water potential in the fluid surrounding cells avoids osmotic problems.

Negative feedback

- Homeostasis is usually achieved by a process called negative feedback.
- Negative feedback is the process in which a movement of a factor away from a set level brings about changes which return it to its original set value.

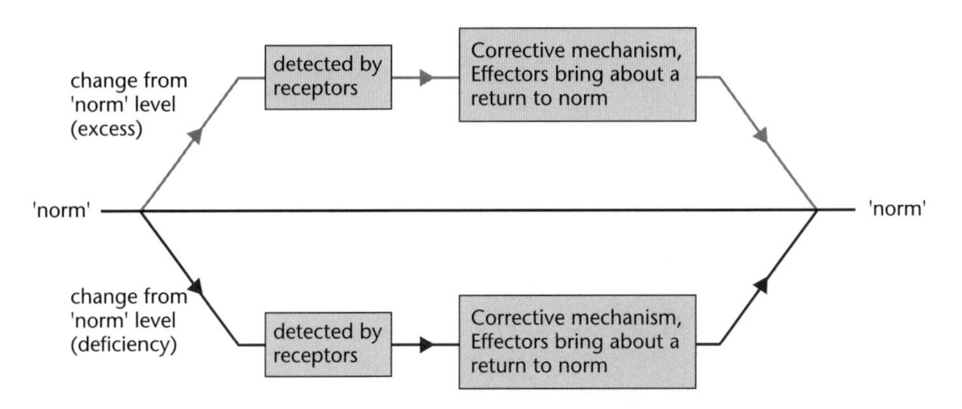

FIGURE 9.1 Negative feedback.

The control of blood glucose concentration

- The **pancreas** contains cells which are sensitive to blood glucose concentration and, thus, act as **receptors**.
- The **islets of Langerhans** within the pancreas produce two different hormones in response to a change in blood glucose concentration.
- These hormones both travel in the blood to their target organs, liver and muscles.
- When the blood glucose concentration is high, **insulin** is produced.
  - It speeds up the rate at which glucose is taken into the cells from the blood.
  - It activates enzymes which are responsible for the conversion of glucose to glycogen in the liver and muscles.
- When the blood glucose concentration is low, **glucagon** is produced.
  - It activates enzymes which are responsible for the conversion of glycogen to glucose in the liver.

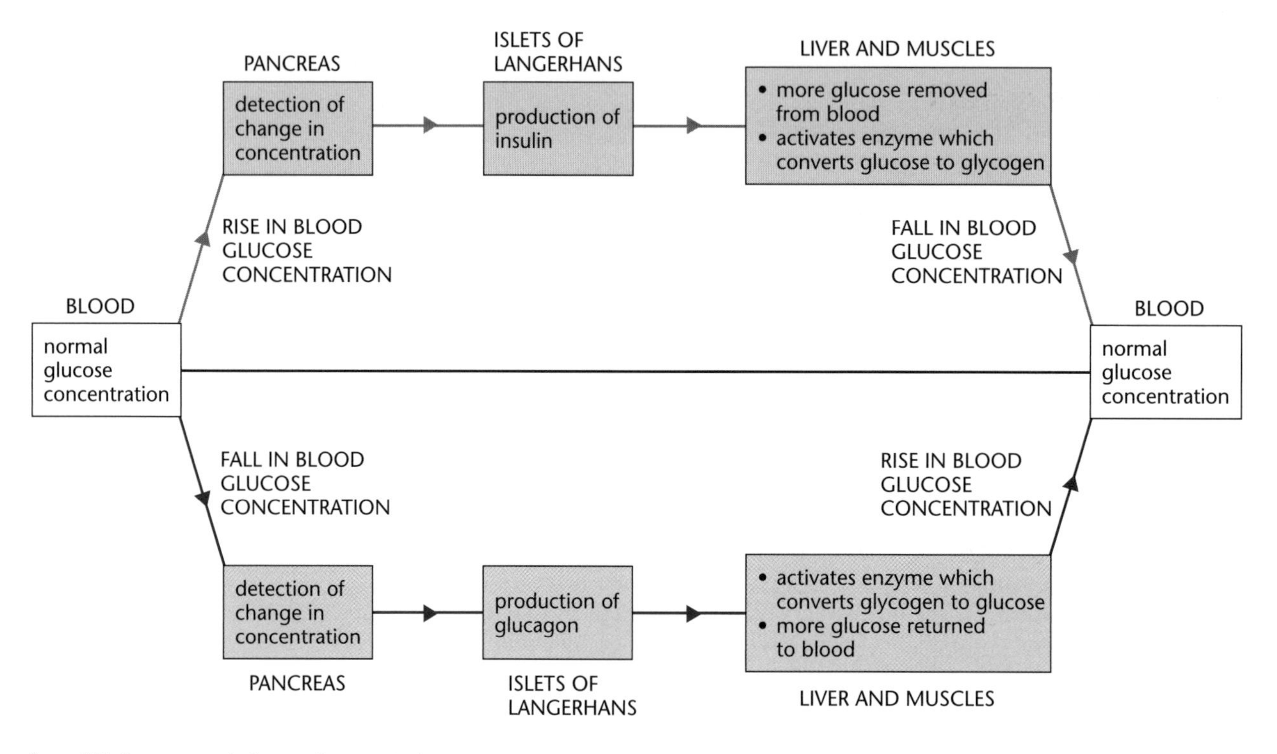

FIGURE 9.2 Summary of glucose homeostasis.

DIABETES

- Early onset diabetes is a condition in which the cells of the islet of Langerhans are **unable to produce insulin**.
- Thus when carbohydrates are eaten and absorbed, the level of blood glucose **rises** dramatically
  - The level rises above the **renal threshold** and glucose is lost in the urine.
  - Because the cells of the kidney cannot reabsorb all the glucose, large volumes of water are lost in the urine, and the person constantly feels thirsty.

- Early onset diabetes can be controlled by:
  - diet, which regulates carbohydrate intake.
  - injections of insulin.

1 If you are not a diabetic, and you eat too much glucose which is not lost in the urine, what happens to it?

You must remember the difference between glucose, glycogen and glucagon.

- Glucose – is a monosaccharide, a sugar found in the blood
- Glycogen – is a polysaccharide which is stored in the liver
- Glucagon – is a hormone which stimulates an increase in blood glucose

Temperature control

All animals keep their internal temperature within certain limits.

- Too high, and there is a risk of **denaturing enzymes**
- Too low, and **biochemical reactions** will be too **slow**

TEMPERATURE REGULATION IN ECTOTHERMS

- Reptiles are examples of ectotherms.
- They have a limited capacity for controlling their internal body temperature and, thus, their **body temperature fluctuates** with the external surroundings.
- They, therefore, rely on **behavioural mechanisms** such as 'sun bathing' and sitting in the shade, or being nocturnal, to maintain a relatively constant internal temperature.

TEMPERATURE REGULATION IN ENDOTHERMS

- Mammals and birds are examples of endotherms.
- They are able to maintain a **constant internal temperature** despite changes in the external temperature.
- They rely on physiological processes, metabolic energy and behaviour to do this.

CONTROLLING BODY TEMPERATURE

- Outside temperature changes only become important when they affect a person's **core temperature**: i.e. the temperature of blood in the internal organs.
- When blood temperature rises or falls, the **hypothalamus** detects the change.
- The hypothalamus responds by sending nervous impulses to appropriate **effectors** which initiate a suitable response.

2 Humans are endotherms. Suggest behavioural methods that you might use to cool down.

SUMMARY OF TEMPERATURE CONTROL

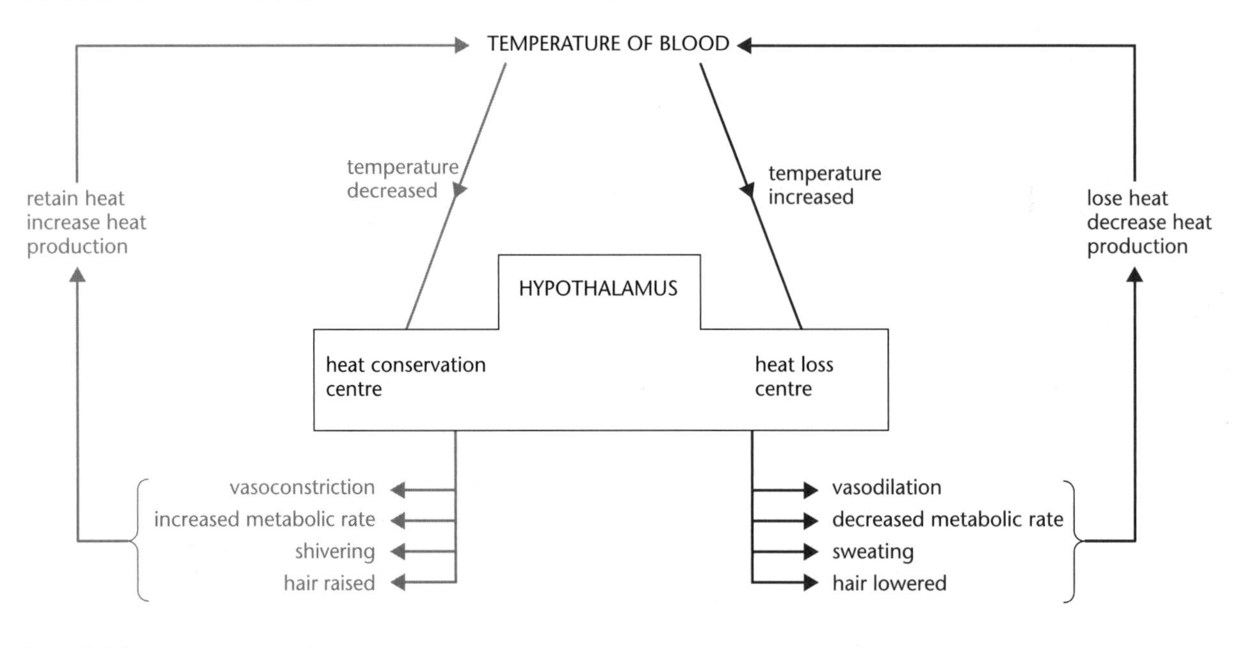

Figure 9. 3 Temperature control.

WORKED EXAM QUESTIONS

1 Mammals are endotherms; reptiles are ectotherms.

a) Explain **two** advantages of endothermy over ectothermy. *(2 marks)*

1 ***Endothermic animals like mammals maintain a constant body temperature regardless of the outside temperature.***

2 ***Endothermic animals can survive in lots of different environments***

Although the candidate has identified a major advantage of endothermy in terms of environmental range, he wasted the first chance to pick up marks by trying to define endothermy, which was not required. The maintenance of a constant internal temperature means that enzymes will function at their optimum rate, and therefore reactions will proceed more quickly. Both those ideas would be worth a mark.

The following questions relate to the information given in two sets of graphs. You will not have been taught this, and may never have seen these graphs. Examiners must include unknown material, so do not expect to recall the information you are given. You will be able to work out the answers from the data you are given and your understanding of this topic. Your experience of working with data during your coursework exercises will also be valuable. Read the questions carefully – answer the question clearly – do not panic!!

b) **Figure 1** shows how the rates of metabolic heat generation and evaporative heat loss in a reptile change with environmental temperature. Each plot is the mean of several values. The vertical bars on the graphs represent the standard deviation about the mean.

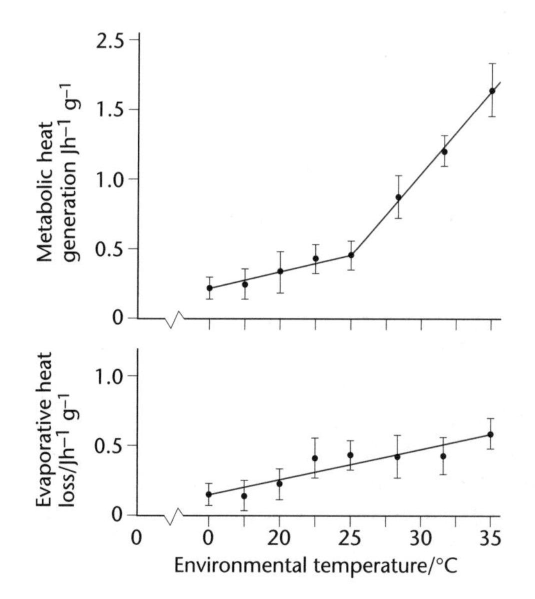

FIGURE 1

(i) Explain why it is more useful to show the standard deviation rather than the range of values. *(2 marks)*

The range will just show the highest and lowest values which may give a false impression if there is a single very high or very low result.

This answer concentrates on one side of the argument, and the examiner cannot be clear that the candidate knows exactly what advantage showing standard deviation has to offer. Standard deviation shows the spread of the 'majority' of the readings, so unlike the range, it will be less affected by anomalous results.

(ii) Explain why the values for metabolic heat generation are given per gram of body mass. *(2 marks)*

If the animals concerned were different weights then a heavier animal might produce more heat than a lighter one even though they were the same.

It is clear that the candidate knows that the principle here is to allow a comparison to be made between animals of different mass (not 'weight' please). Greater mass may lead to an increase in heat generation, even if the metabolic rate of the two organisms is the same. However, he finds it hard to express those ideas.

(iii) Describe the relationship between metabolic heat generation and evaporative heat loss shown in the graphs. *(2 marks)*

Evaporative heat loss continues to rise as environmental temperature rises. Although metabolic heat generation rises with environmental temperature up to 25°C it then increases faster than it did before.

When asked to describe trends in graphs, always refer to the variables, and, if possible, state the values where changes in the slope of the graphs occur. However, this question asks for a comparison between the amount of heat generated, and that lost by evaporation. The patterns shown by the two graphs need to be compared, but the candidate simply describes the relationship between the independent and dependent variables on the two graphs separately. The correct answer was that both increase proportionally up to 25°C, but then the heat generated increases faster than the heat lost by evaporation.

(iv) Use the graphs to explain why these reptiles often seek shade when the environmental temperature rises above 25 °C. *(2 marks)*

Up to 25°C the reptile can lose as much heat as it produces by evaporation but at temperatures greater than 25°C it over heats and would die if it did not get out of the sun.

Again, the basic idea is here, but the answer is poorly written. The sentence construction suggests that the candidate thinks that the reptile produces heat by evaporation. However, he has realised that the reptile will overheat, which is worth one mark; but he should have explained that reptiles have no physiological cooling mechanism. They are unable to vasodilate, sweat or reduce the insulating properties of their skin surface by flattening hairs.

c) **Figure 2** shows the relationship between metabolic heat generation and evaporative heat loss in a small mammal.

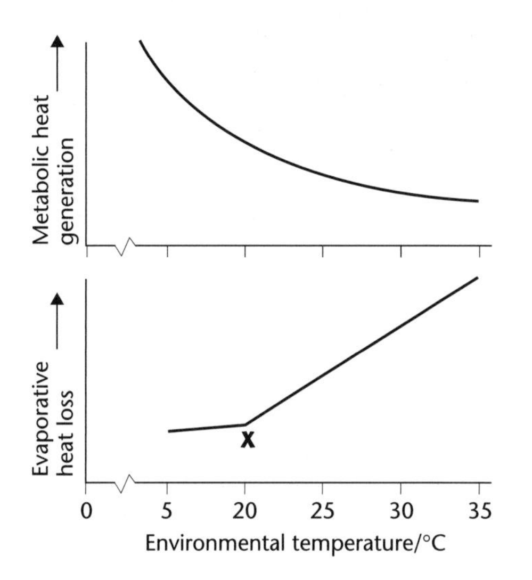

FIGURE 2

(i) How is the relationship between heat generation and evaporative heat loss in a mammal different from that in a reptile? *(1 mark)*

The relationship is the opposite.

This answer would just get the mark, but the candidate should explain clearly what he means. It this case, when environmental temperatures are greater than 20°C, the rate of heat lost by evaporation increases, whilst the amount of heat produced by metabolism falls. This is the reverse of the situation found in a reptile, where the heat produced by metabolism increases rapidly at higher environmental temperatures, but the heat lost by evaporation rises only slowly.

(ii) Suggest an explanation for the change in the slope of the graph for evaporative heat loss at the point marked **X**. *(1 mark)*

More heat is lost by evaporation as the mammal is sweating.

It is not clear whether the mammal is sweating all the time and the rate simply increases at this point, or whether sweating begins in addition to other methods. However, the candidate has identified the process of sweating and would get the mark.

(iii) Explain how the change in metabolic heat generation in a small mammal is brought about as environmental temperature rises. *(3 marks)*

Less heat is generated because the animal does not move about so much in hotter weather. This means that it does not breathe so fast and needs less energy to move its muscles. When muscles move they need ATP but the process of respiration is not very efficient and lots of energy from glucose is lost as heat.

Less physical activity and reduced respiration are valid answers; although again the candidate has taken a long time to write that. Another reason for the fall of metabolic heat is the reduced activity of organs, like the liver, and the reduced secretion of hormones, like thyroxine, which speed up metabolic reactions in cells.

(AQA 2003)

EXAMINATION QUESTIONS

1 a) Humans are able to maintain a constant core temperature when exposed to cold external temperatures.

Suggest

(i) **one** advantage of this; *(1 mark)*

(ii) **one** disadvantage of this. *(1 mark)*

b) The graphs show data collected from a volunteer who ate several ice cubes.

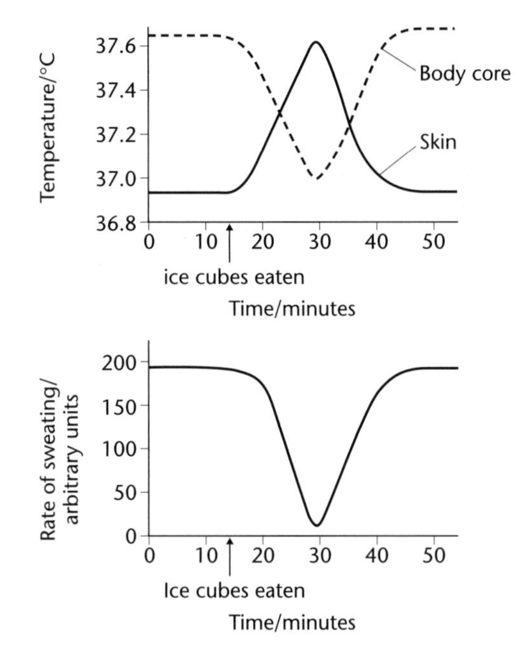

(i) Explain the relationship between the rate of sweating and the temperature of the skin. *(2 marks)*

(ii) Use information in the graphs to explain the part played by negative feedback in the control of core temperature. *(2 marks)*

(AQA 2003)

10 Gas exchange

After revising this topic, you should:

- understand that that all organisms rely on **diffusion** for the exchange of respiratory gases
- be able to explain how the **efficiency of gas exchange** is increased by:
  - a large surface area
  - a large difference in concentration across the gas exchange surface
  - a thin gas exchange surface
- be able to describe the relationship between the **size** of an organism and its **surface area to volume ratio**; and explain the importance of this in the process of gas exchange
- understand that gas exchange in terrestrial organisms always involves a loss of water from the exchange surface
- be able to describe the main features of the gas exchange surface in a protoctist, an insect, a fish, and a plant.

The factors determining the rate of diffusion

$$\text{Diffusion rate} \propto \frac{\text{surface area} \times \text{difference in concentration}}{\text{thickness of exchange surface}}$$

The most efficient gas exchange surface will be the one through which the rate of diffusion will be as high as possible, so it will have adaptations which:

- provide a large surface area
- maintain a large difference in concentration
- ensure that the exchange surface is as thin as possible

Size and surface area to volume ratio

Diffusion depends on surface area, so a very small organism such as an amoeba, which has a large surface area to volume ratio, can meet all its gas exchange requirements by diffusion through its surface. Larger organisms, such as insects, fish and mammals, have a relatively small surface area to volume ratio. They have **specialised gas exchange systems** that increase the surface over which diffusion can take place.

?

1 Imagine three animals, all cube shaped, one having 1 cm length sides, one having 2 cm length sides and one having 3 cm length sides.
a) Calculate the surface area to volume ratio for each.
b) Explain one advantage and one disadvantage of being small.

- Remember, any system involved in concentration gradients, such as the counter-current system in gills, helps to **maintain** a concentration difference. It is not enough to say that it is large.
- In the counter-current system, the blood and the water move in **opposite directions** maintaining the concentration difference across the whole of the gill.

2 Explain under what circumstances plants take in oxygen.

Limiting water loss

Surfaces of terrestrial organisms have modified cells which limit water loss. These cells, however, also prevent the exchange of gases. Thus, when gas exchange occurs, water is bound to be lost. Water loss is a major problem and gas exchange systems are modified to reduce it.

PLANTS

A waterproof cuticle covers the leaf, but water will escape from the cells to the atmosphere. This water loss is reduced if:

- the exchange surface is inside the leaf
- the stomata can be closed

There is potential conflict between limiting water loss and obtaining carbon dioxide through the stomata for photosynthesis. The extent of opening is a compromise.

INSECTS

A waterproof cuticle covers the insect's body, but water will escape from the cells to the atmosphere. This water loss is reduced if:

- the spiracles can be closed
- 'hairs' around the spiracle can trap moist air

There is potential conflict between limiting water loss and obtaining oxygen and removing carbon dioxide through the spiracles. Again, the extent of opening is a compromise.

Transpiration is not sweating. In a leaf, water evaporates into the air spaces then leaves through the stomata down a water potential gradient. When you sweat, liquid water is secreted onto the surface of your skin.

Gas exchange surfaces

	Gas exchange surface in:		
	Protoctist	**Fish**	**Insect**
Diagram of exchange surface	Cell surface membrane 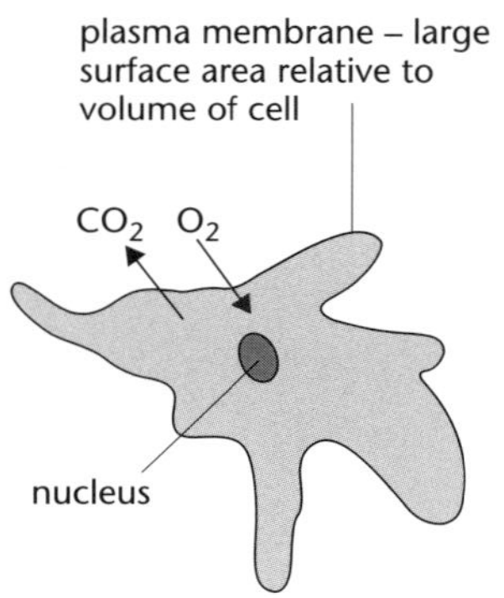 plasma membrane – large surface area relative to volume of cell CO_2 O_2 nucleus	Gill lamellae blood vessels gill plates lamellae gill plate lamella deoxygenated blood – from the heart water flow blood flow opposite to water flow oxygenated blood – going to tissues 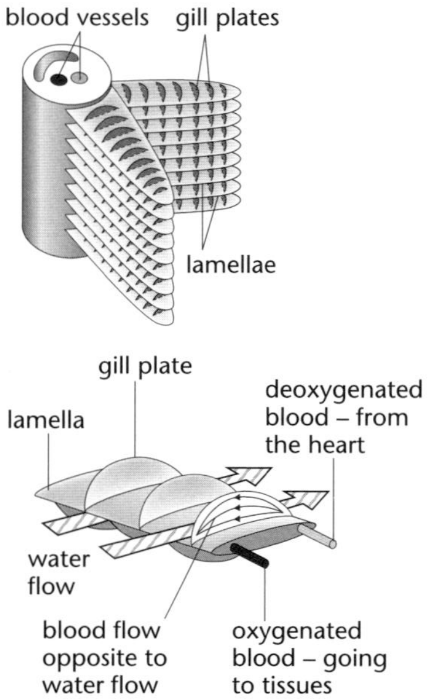	Tracheoles outer surface spiracle tracheae tracheole O_2 body cell CO_2 rings of chitin 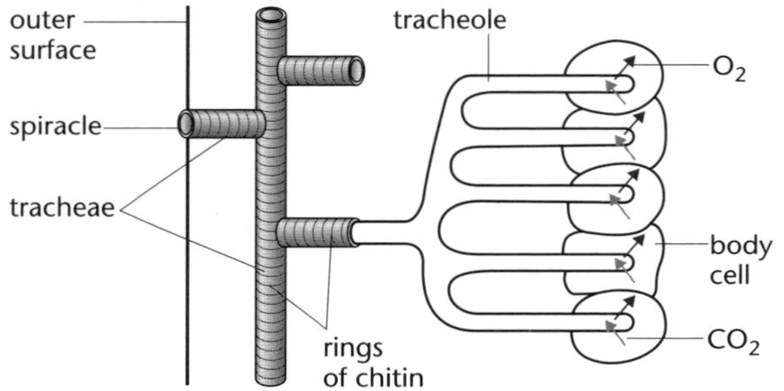
Ventilation	None	Movements of buccal cavity and operculum create a one way flow of water	Abdomen dilates, which decreases pressure and draws air in
Large surface area	Small size, therefore a large surface area to volume ratio	Many lamellae covered with many gill plates	Large number of tracheoles
Large concentration difference maintained	Oxygen is used by the cell: therefore its concentration is always less in the cell than outside	◆ Oxygenated water always moving across the gills ◆ Counter-current system in lamellae ◆ Circulation of blood removes the oxygen	Oxygen is used by the cells of the insect

Thin exchange surface	Plasma membrane only 7 nm: therefore very short diffusion pathway	Only two complete layers of cells between the water surrounding the gills and the blood in the capillaries	Thin wall of tracheole

	Gas exchange surface in:	
	Spongy mesophyll	**Mammal**
Diagram of exchange surface	Plasma membrane of spongy mesophyll cells cuticle; upper epidermis; SPONGY MESOPHYLL This is a layer of small cells with large irregular air spaces; VASCULAR BUNDLE; guard cell; stoma; lower epidermis; large intercellular air space	Alveoli bronchiole; alveoli; capillary network around alveolar walls
Ventilation	None	Movements of diaphragm and ribs decrease pressure which draws air in
Large surface	Many cells in contact with the air spaces inside the leaf	Large numbers of alveoli
Large concentration difference maintained	Carbon dioxide is used by the mesophyll cells for photosynthesis	◆ Oxygenated air is always entering the alveoli ◆ Circulation of blood continually removes the oxygen
Thin exchange surface	Thin cellulose cell wall and plasma membrane between cells and air spaces	◆ Capillaries close to alveoli ◆ Both structures are made of one layer of squamous epithelia ◆ Only two layers of cells between the air in the alveolus and the blood in the capillaries

WORKED EXAM QUESTIONS

1 The drawing shows some tracheoles that carry air to a muscle fibre in an insect.

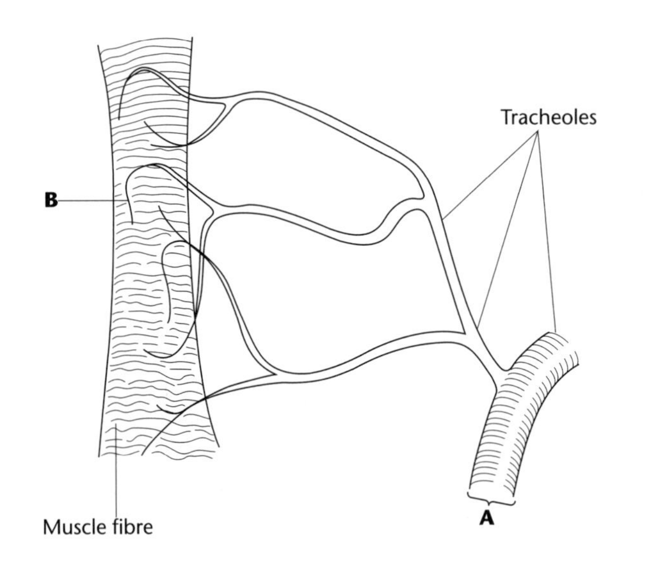

a) (i) Explain how the structure of the gas exchange system of an insect ensures that there is a large surface area for gas exchange. *(1 mark)*

An insects gas exchange system is made up of many tubes that link with the outside through spiracles

> Although the candidate has described the tracheal system, she has not really addressed the question. The large surface area is achieved by the tracheal system because it is made up of lots of small, **branched** tracheoles.

(ii) Describe **one** way in which the transport of oxygen to a muscle in an insect is different from that in a fish. *(1 mark)*

Oxygen travels directly to the muscle in an insect from the tracheoles, it does not go into the blood first.

> This is a good answer, it is clear and to the point. The branching tracheal system of an insect carries oxygen close to every cell. The blood of an insect has no respiratory pigment (like haemoglobin) but does have similar roles in terms of transport of nutrients, hormones and metabolic waste. It contains white blood cells too, and has a role in an immune response.

S b) The diameter of the tracheole at point **A** is 20 μm. Calculate the magnification of the drawing. Show your working. *(2 marks)*

300 times

S, synoptic from module 1. Although this candidate has not shown any working, she has the correct answer and therefore was given both marks. The reason why most questions that involve calculations ask for the workings to be shown is to allow examiners to search for credit if the answer is incorrect. Candidates may have used the correct method but may have failed to do the calculation correctly; we do try hard to find evidence to award some of the available marks.

The diameter of the tracheole at point **A** on the diagram must be measured carefully. It is 6 mm. Convert that distance to the same units given in the question (μm).
6 mm = 6 000 μm and then divide by the actual length to find the magnification.

$$\frac{6\,000}{20} = 300$$

S c) Breathing movements can bring about the mass flow of air as far as point **B**. What causes the diffusion of oxygen molecules from **B** into a muscle fibre? *(1 mark)*

Diffusion takes place rapidly because the surface area is large, and the distance to travel is very small. There is also a difference in the concentration of oxygen in the tracheal and the muscle. Diffusion is the passive movement of a gas from a high to a low concentration.

S, again from module 1. This candidate has seen the word 'diffusion' and has correctly thought back to Fick's law from module 1. However, the question expects candidates to be selective. It asks 'what causes diffusion' not what makes diffusion occur most rapidly. So the question expects a reason for the difference in concentration that the candidate describes. The correct answer is that the muscle uses oxygen in aerobic respiration, and, therefore, oxygen concentration in the muscle is always lower than the oxygen concentration in the tracheoles.

(AQA 2003)

EXAMINATION QUESTIONS

1 a) A diffusion gradient is essential for gas exchange. Describe **two** ways in which a diffusion gradient for oxygen is maintained at the gas exchange surface of a fish. *(2 marks)*

The graph shows the concentration of gases inside the tracheoles of an insect.

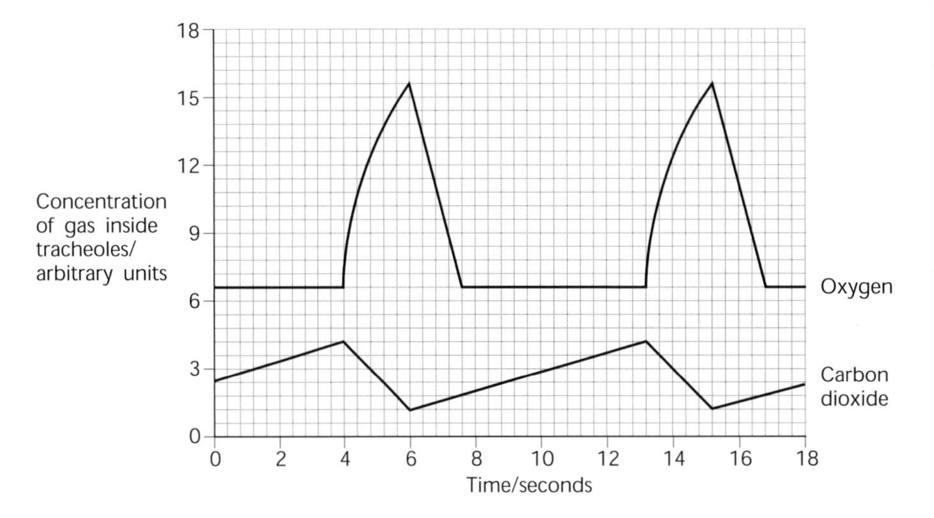

b) For how long were the insect's spiracles open during the period shown in the graph? Explain how you arrived at your answer.

Length of time:

Explanation:

(2 marks)

(AQA 2002)

11 Transporting respiratory gases in mammals

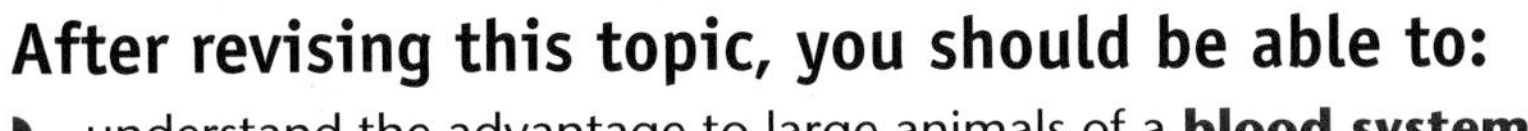

After revising this topic, you should be able to:

- understand the advantage to large animals of a **blood system** that transports respiratory gases
- describe the role of **haemoglobin** in transporting oxygen from the lungs of a mammal to its respiring tissues
- explain what is meant by the **Bohr shift** and describe its importance in increasing the oxygen made available to respiring tissues
- relate the oxygen-carrying properties of **different types of haemoglobin** to the environment in which the animal lives
- describe how **carbon dioxide** is transported in the blood

Introduction

Oxygen and carbon dioxide are **respiratory gases**. Oxygen is required by every cell for aerobic respiration. Carbon dioxide is a waste product of aerobic respiration and has to be removed from the body.

- Oxygen passes through gas exchange surfaces and enters respiring cells by simple diffusion.
- Diffusion is efficient if the distances involved are very small.
- However, in large organisms, each cell is too far from the oxygen supply – they have blood systems, **carrying oxygen**, close to every cell. This aids the process of diffusion, and ensures that all cells in the organism receive an adequate supply of oxygen.

> **?** **1** Why must carbon dioxide be removed from the body?

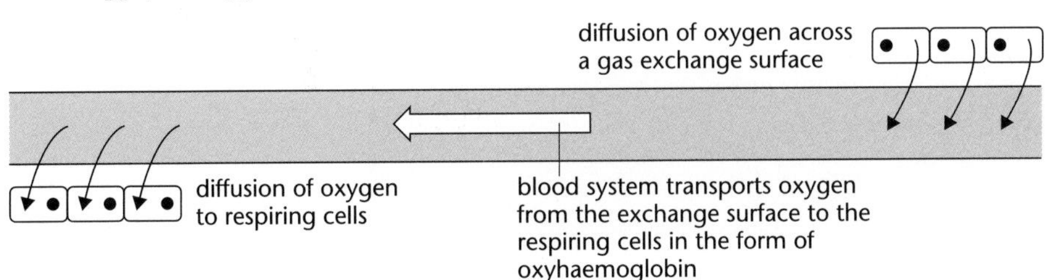

FIGURE 11.1 Oxygen supply to cells of a large organism.

The role of haemoglobin

- Oxygen is carried in red blood cells combined to the protein, haemoglobin.
- Haemoglobin binds (associates) with oxygen in the lungs, where oxygen is plentiful, forming **oxyhaemoglobin**.
- Oxyhaemoglobin releases (dissociates from) oxygen to the respiring cells. It gives up oxygen which diffuses into the cells while the haemoglobin remains in the red blood cells and is transported in the blood back to the lungs.

DISSOCIATION CURVE

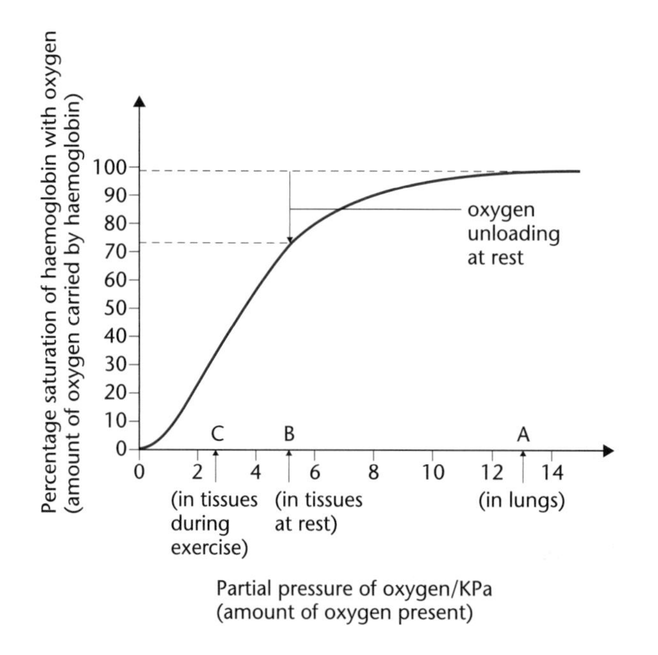

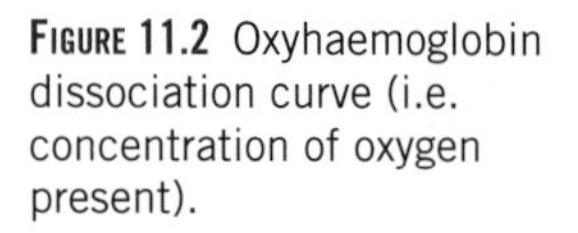

FIGURE 11.2 Oxyhaemoglobin dissociation curve (i.e. concentration of oxygen present).

The graph shows how much oxygen the haemoglobin is carrying depending on the concentration of oxygen present.

Point A	Point B
◆ This point represents the partial pressure of oxygen in the lungs ◆ The haemoglobin is 98% saturated	◆ This point represents the partial pressure of oxygen in respiring tissues at rest ◆ The haemoglobin is 73% saturated
◆ Therefore, oxygen that was being carried by haemoglobin at A has been given up at B ◆ In this example 98 – 73 = 25 so 25% of oxygen has been released	

2 Look at point C on the graph. This point represents the partial pressure of oxygen in respiring tissues during exercise.

a) How much oxygen is unloaded?

b) Explain the advantage of this in these tissues.

EXAMINER'S TIP

- **Partial pressure of oxygen** represents the concentration of oxygen available.
- **Percentage saturation** represents the amount of oxygen being carried by oxyhaemoglobin.

Be sure you understand what the terms mean.

The Bohr Effect

In the body, the ability of haemoglobin to transport oxygen is also affected by the partial pressure (concentration) of **carbon dioxide** present.

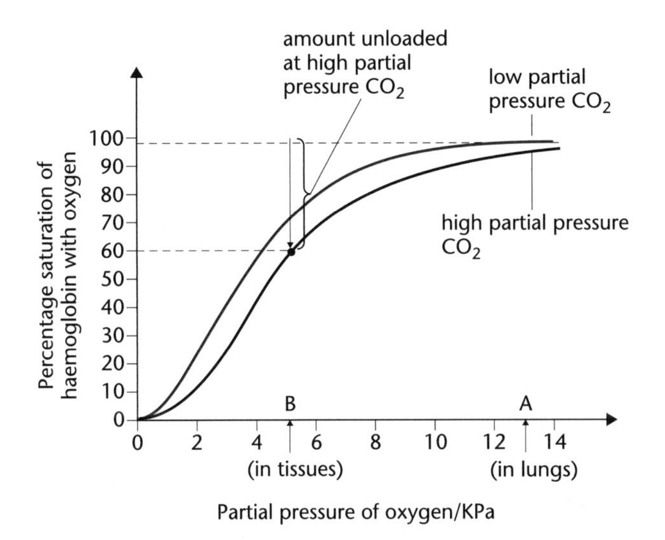

FIGURE 11.3 Diagram of Bohr effect.

The Bohr effect:

- occurs when carbon dioxide concentration is increased
- always has the effect of displacing the dissociation curve to the right.

Point A	Point B
◆ Represents the partial pressure of oxygen in the lungs ◆ There are two lines: one representing high partial pressure of carbon dioxide (black) and the other low partial pressure of carbon dioxide (red) ◆ In the lungs there will always be a low partial pressure of carbon dioxide, so we use the red curve ◆ The red curve shows that the haemoglobin is 98% saturated	◆ Represents the partial pressure of oxygen in respiring tissues ◆ There are two lines: one representing high partial pressure of carbon dioxide (black); and the other low partial pressure of carbon dioxide (red) ◆ Since the cells are respiring, they will be using up oxygen and releasing carbon dioxide as a waste product, so there will be a high partial pressure of carbon dioxide ◆ So, using the red curve, the haemoglobin will be 60% saturated
◆ The effect of increasing the partial pressure of carbon dioxide results in the oxyhaemoglobin giving up **even more** of the oxygen it is carrying to the tissues. ◆ In this example, 98 – 60 = 38. So 38% of oxygen has been released. This is 13% more than if the low partial pressure of carbon dioxide in the lungs were maintained.	

SUMMARY

When the rate of respiration increases, oxyhaemoglobin releases more oxygen to the respiring cells. There are two reasons for this.

- The greater the rate of respiration, the more oxygen a cell consumes. This results in a lower partial pressure of oxygen, which increases the amount of oxygen that the oxyhaemoglobin will release. This is shown in Figure 11.2: as the partial pressure of oxygen falls, so does the percentage saturation of the oxyhaemoglobin.
- The greater the rate of respiration, the greater the amount of carbon dioxide that is produced. This pushes the dissociation curve to the right: again encouraging the oxyhaemoglobin to release more of the oxygen it is carrying. This is shown in Figure 11.3.

3 Haemoglobin is a buffer.
a) What is the meaning of the term buffer?
b) Explain why it is important to have a buffer in the blood.

Different sorts of haemoglobin

Different organisms have different sorts of haemoglobin, allowing them to live successfully in different environments.

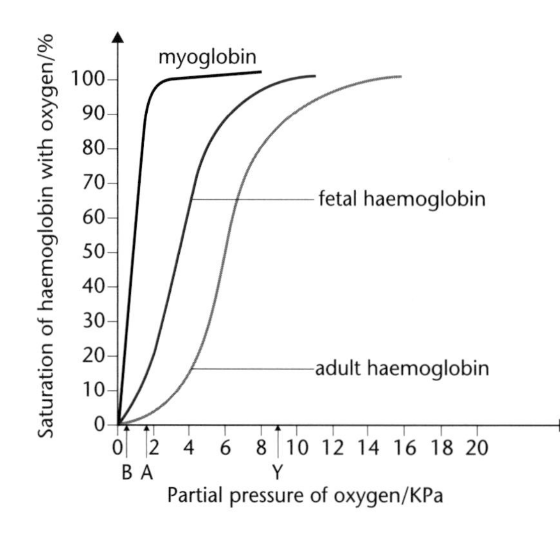

FIGURE 11.4 Dissociation curves of myoglobin, fetal haemoglobin and adult haemoglobin.

Myoglobin	Fetal haemoglobin
◆ This is always represented as a curve far to the left, with a 'dogleg' shape rather than an 'S' shape. ◆ At very low partial pressures of oxygen, point A, the myoglobin is fully saturated.	◆ This has the same 'S' shape as adult haemoglobin, but it is always to the left of the curve for adult haemoglobin. ◆ This pigment becomes fully saturated at lower partial pressures of oxygen; see point Y on the graph.

◆ As the partial pressure of oxygen falls to a very low value, point B, myoglobin readily releases all the oxygen it is carrying. ◆ Thus myoglobin acts as an oxygen store. ◆ It is found in the blood of animals such as seals and in our muscles.	◆ The position of this curve (when compared with adult human haemoglobin) is typical for any organism that is **living in an environment with a low partial pressure of oxygen**. ◆ Fetuses receive oxygen from the placenta, which has a lower partial pressure of oxygen than the lungs of the mother. It is vital that the fetal haemoglobin becomes fully saturated to supply sufficient oxygen to growing tissues. ◆ There are many more examples: – llamas live at high altitude, and, therefore, have lower partial pressures of oxygen – burrowing mammals

Getting rid of carbon dioxide

Carbon dioxide is a waste product of respiration. It is taken to the lungs where it is removed from the body.

Carbon dioxide is transported in the blood to the lungs in a number of ways

- ◆ Carbon dioxide can dissolve in water. About 5% of cabon dioxide in blood is transported **in solution** in the plasma. Some slowly forms hydrogen carbonate ions.
- ◆ Carbon dioxide reacts directly with haemoglobin to form **carbamino-haemoglobin**. The amount that haemoglobin can carry in this way depends on the amount of oxygen it is carrying but is usually about 10%. The smaller the amount of oxygen, the more carbon dioxide it can carry.
- ◆ Most carbon dioxide (85%) is transported as **hydrogencarbonate ions** (HCO_3^-) in the plasma.

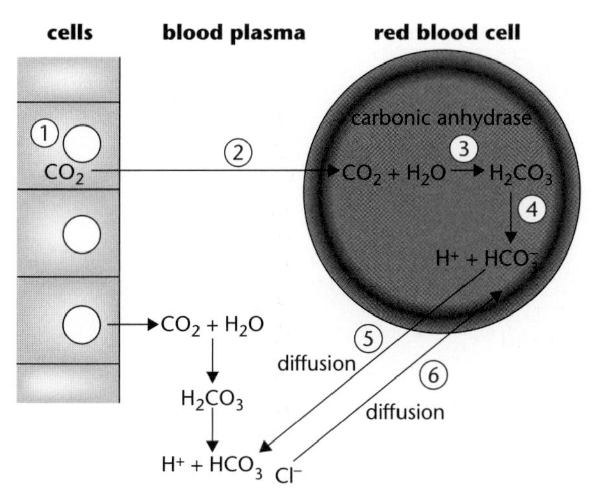

FIGURE 11.5 Carbon dioxide transport.

1 Carbon dioxide is produced in respiring cells.
2 CO_2 diffuses from the cells through the plasma into the red blood cells.
3 The carbon dioxide reacts with water to form carbonic acid (H_2CO_3). The enzyme, carbonic anhydrase, inside the red blood cell, speeds up this reaction.
4 The carbonic acid dissociates to form hydrogen ions (H^+) and hydrogencarbonate ions (HCO_3^-).
5 As there is a high concentration of hydrogencarbonate ions in the red blood cell, they diffuse out of the red blood cell into the plasma down a concentration gradient.
6 A loss of hydrogencarbonate ions causes an electrochemical gradient, which results in chloride ions (Cl^-) moving into the cell. This is called the **chloride shift**.

WORKED EXAM QUESTIONS

1 The graph shows the dissociation curve for adult human haemoglobin.

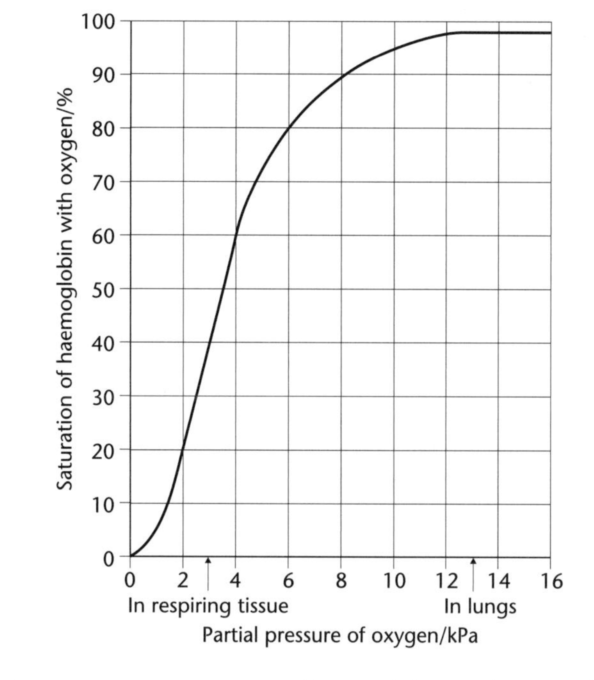

a) What is the percentage saturation of haemoglobin at the partial pressure of oxygen found in

(i) the lungs;

(ii) respiring tissue? *(1 mark)*

i) 98%

ii) 40%

Read the information given on the graph to be sure you know exactly what the examiner wants you to do. In this case, you are not expected to know the partial pressure of the lungs or of respiring tissue; those values are given on the graph. When taking readings from a graph, do not be afraid of drawing on the exam paper. Take a ruler and draw a line from the *x*-axis to meet the graph and then from that point to the *y*-axis. Both these answers are correct.

b) 1 dm^3 blood leaving the lungs carries 200 cm^3 oxygen. Use the graph to calculate the amount of oxygen that this volume of blood will unload to the respiring tissue. Show your working. *(2 marks)*

In the lungs, blood is 98% saturated and this carries 200 cm^3 oxygen
100%-saturated blood would carry 204 cm^3 oxygen
1%-saturated blood would carry 2.04 cm^3 oxygen
In the respiring tissues, blood is 40% saturated so it would carry
$40 \times 2.04 = 81.60$ cm^3 oxygen

Although the basic calculation has been done correctly, this candidate has not read the question carefully enough. He has calculated the amount of oxygen carried by the blood in respiring tissue, not the amount that was unloaded. He needed to subtract the percentage saturation in the tissues from that in the lungs (98 – 40 = 58). This would have given him one of the two marks available, and is why the question asks the candidate to show his working.
The difference gives the amount of oxygen unloaded. $58 \times 2.04 = 118.32\ cm^3$.
Check any mathematical answer using common sense. In this case, just over half of the oxygen, 58%, is released so as 118 is slightly more than half of 200 it looks right.

S c) A mutation in one of the genes coding for the production of haemoglobin could lead to a decrease in the oxygen-carrying capacity of the blood. Explain how. *(2 marks)*

A mutation in a gene could be an addition, a deletion or a substitution. The first two would lead to a frame shift and therefore all the codes for amino acids following that mutation would change. A completely different protein would be produced which may not be able to carry oxygen at all. If a substitution took place only one of the codes would alter and therefore only one different amino acid would be coded for. The protein in haemoglobin may be a little different but may still be able to carry some oxygen.

S indicates a synoptic question, this time from module 2. This candidate obviously knows some of the terms, but is determined to give the examiner all he knows, rather than answer the question. He has continued to write in the space under the lines provided, which should be unnecessary. However, he has omitted some important ideas. The terms addition, deletion and substitution are not qualified. What had been added? He never makes it clear that it is a change of one of the bases or nucleotides of DNA that is referred to as a mutation. He has however linked mutation with a change of the protein structure of haemoglobin, and, despite not focusing on the question, will be awarded one mark.

(AQA 2003)

EXAMINATION QUESTIONS

1 The llama is a mammal which lives at high altitude. Sheep live at low altitude. The graph shows dissociation curves for llama and sheep haemoglobin.

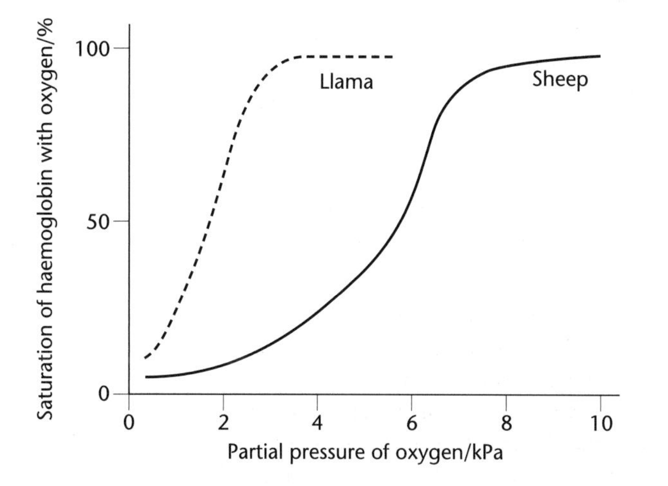

a) Explain the advantage of the shape and position of the llama's dissociation curve. *(2 marks)*

b) Haemoglobin acts as a buffer, preventing changes in blood pH. Such changes could lead to a decrease in the rate of enzyme-controlled reactions in the body.

(i) Explain how haemoglobin is able to act as a buffer. *(1 mark)*

S (ii) Suggest how changes in pH could lead to a decrease in the rate of enzyme-controlled reactions. *(1 mark)*

(AQA 2002)

12 Digestion and diet

After revising this topic, you should be able to:

- identify the different stages in the processing of food in the gut of a mammal
- describe the role of **digestive enzymes** in the digestion of carbohydrates, lipids proteins and cellulose
- explain how the structure of the small intestine is adapted to its functions of digestion and absorption
- explain how the products of digestion are **absorbed**
- explain how the **secretion** of digestive enzymes is controlled
- describe the life cycle of a **lepidopterous insect**
- explain the change in protein and energy requirements associated with growth in the larva and reproduction and dispersal in the adult
- explain the changes in food and gut enzymes during the insect's life

Introduction

There are four main stages involved in the processing of food in the gut of a mammal.

Stage	Definition
Ingestion	Food is taken through the mouth into the **buccal cavity**
Digestion	Large insoluble molecules (which make up the food of an organism) are broken down to smaller, soluble molecules by the action of various enzymes
Absorption	The products of digestion are taken into the cells of the body through the **gut wall**
Egestion	The removal of undigested and indigestible food, dead cells and bacteria as **faeces**

The role of digestive enzymes

STARCH DIGESTION

Part of gut	Secretion	Enzyme	Substrate	Product	Notes
buccal cavity	Saliva	Amylase	Starch	Maltose	Optimum pH is 7
stomach	No starch digestion takes place here				
duodenum	Pancreatic juice	Amylase	Starch	Maltose	Optimum pH is 8
ileum		Maltase	Maltose	Glucose	The enzyme that catalyses this reaction is located in the plasma membrane of the epithelial cells, which line the small intestine.

1 Starch is a storage molecule found only in plants.
a) Name the carbohydrate storage molecule found in animals.
b) In which organs is this molecule stored in animals?

PROTEIN DIGESTION

Part of gut	Secretion	Enzyme	Substrate	Product	Notes
buccal cavity	No protein digestion takes place here				
stomach	Gastric juice	Pepsin (endopeptidase)	Polypeptides	Peptides	◆ Pepsin is secreted in the inactive form pepsinogen. ◆ It is converted to its active form pepsin by hydrochloric acid.
duodenum	Pancreatic juice	Trypsin (endopeptidase)	Polypeptides	Peptides	◆ Trypsin is secreted in the inactive form trypsinogen. ◆ It is converted to its active form trypsin by the enzyme enterokinase.
		Exopeptidase	Polypeptides and peptides	Amino acids and dipeptides	
ileum		Dipeptidase	Dipeptides	Amino acids	The enzyme that catalyses this reaction is found in the epithelial cells of the small intestine.

EXAMINER'S TIP

- Endopeptidases are released from the cells in an inactive form. They are activated in the gut.
- Endopeptidases break peptide bonds between specific amino acids in the **middle** of polypeptide chains.
- Exopeptidases break peptide bonds between amino acids at the **ends** of polypeptide and peptide chains.

2 Name the process that breaks two amino acids apart by the enzymic addition of water.

LIPID DIGESTION

Part of gut	Secretion	Enzyme	Substrate	Product	Notes
buccal cavity	No lipid digestion takes place here				
stomach	No lipid digestion takes place here				
duodenum	Pancreatic juice	Lipase	Triglycerides	Glycerol and fatty acids	Bile contains bile salts which emulsify the triglycerides.
ileum	No lipid digestion takes place here				

3 What is the difference between a triglyceride and a phospholipid?

RUMINANT DIGESTION

Many mammals feed entirely on plants. The most abundant carbohydrate in their food is cellulose, but they do not produce cellulase enzymes. They rely on a **mutualistic relationship** with bacteria living in their guts.

Ruminants and cellulose

Cellulose →(*cellulase*) cellobiose (disaccharide) → glucose → pyruvate → fatty acids

- Cellulose is a polysaccharide.
- It is broken down to glucose with the help of the bacterial cellulase.
- The glucose formed is used mainly by the microorganisms.
- The glucose is converted to pyruvate.
- The pyruvate is further metabolised producing:
  - waste products such as carbon dioxide and methane, which are lost
  - waste products such as fatty acids, which are absorbed through the wall of the rumen into the blood of the cow.

4 What is the difference between the monomers that make up starch and cellulose?

EXAMINER'S TIP

- Bile does NOT contain enzymes.
- Bile increases the surface area of the triglycerides for the action of lipase to be more efficient.
- Sodium hydrogen-carbonate, (secreted from the pancreas and in bile from the liver) provides the optimum, slightly alkaline, pH for the enzymes.

Ruminants and protein

The microorganisms living in the rumen also play an important role in **providing protein** for the ruminant to supplement its protein-poor diet.

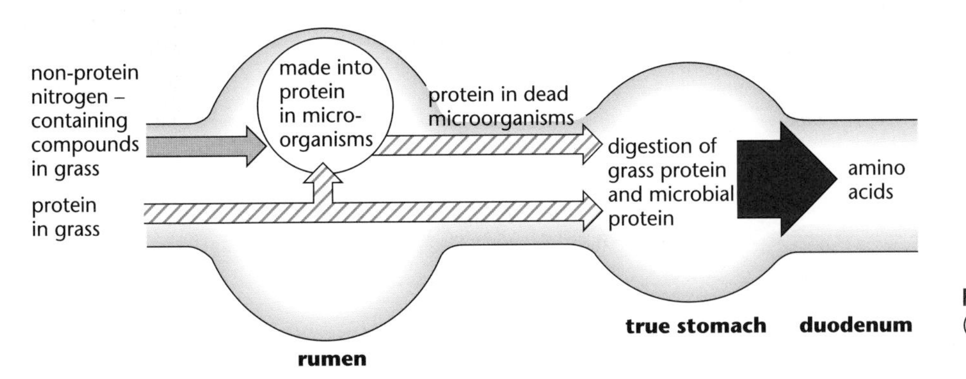

Figure 12.1 Digestion in cattle (an example of a ruminant).

- Most of the protein in the grass is digested by the ruminant to form amino acids.
- Non-protein nitrogen-containing compounds e.g. DNA, are used by the microorganisms in the rumen to make microbial protein.
- When the microorganisms die, the protein in their cells is digested by the ruminant.

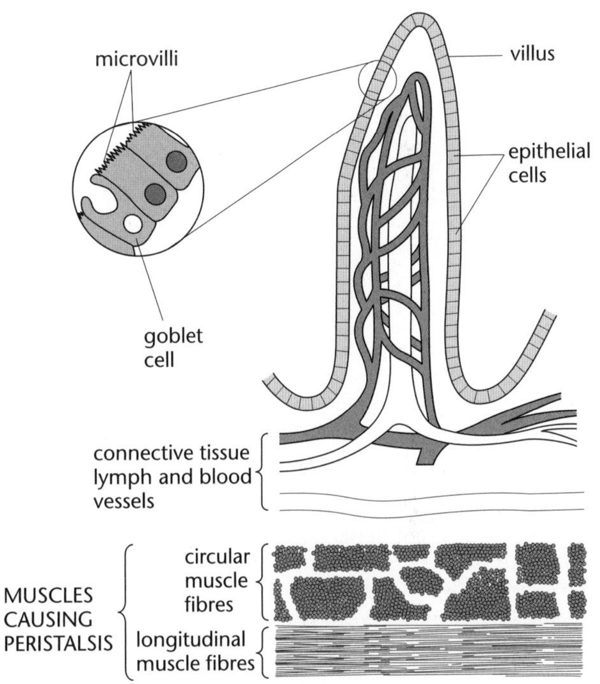

Figure 12.2 Structure of a villus.

The small intestine

The ileum of the **small intestine** is the region of the gut where most of the digested food is absorbed.

Food moves along the small intestine by the action of **peristalsis**.

The wall of the small intestine forms a very efficient absorption surface because it is specially adapted, see table below.

Feature	Notes
Large surface area over which diffusion and active transport can take place	◆ It is long ◆ It is folded and the folds are covered in **villi** ◆ The epithelial cells that form the lining of the villi have plasma membranes that are further folded into **microvilli** ◆ Many carrier molecules involved in active transport are involved in absorption
It maintains a **large difference in concentration**	◆ Muscles in the gut wall ensure that the contents of the small intestine are moving constantly ◆ The circulation of blood in the capillaries of the villi ensures that the absorbed amino acids and glucose are removed quickly
Thin exchange surface	◆ The **epithelial cells** give a short distance between the lumen of the gut and the blood

Absorbing digested food

Substance to be absorbed	Diagram of process	Notes
Salts and water	epithelium; capillary; lumen of ileum; microvilli; Na^+; Na^+; Na^+; ATP; ADP; sodium ion channel protein in plasma membrane of microvillus; sodium ion pump	◆ Sodium ions pass into the epithelial cells by diffusion through ion channel proteins ◆ They are actively transported from the epithelial cells into the blood ◆ Active transport maintains a low concentration of sodium ions in the cell ◆ This allows continuous uptake from lumen of gut
Glucose	epithelium; capillary; lumen of ileum; Na^+; glucose; Na^+; ATP; ADP; glucose; co-transport protein in plasma membrane of microvillus; glucose transport protein if the sodium ion concentration in the epithelial cell becomes too high, the co-transport proteins will not take up as much glucose	◆ Glucose is absorbed from the gut by diffusion ◆ This is coupled with the absorption of sodium ions ◆ There is a specific carrier molecule that takes both Na^+ ions and glucose molecules into the cytoplasm together ◆ Na^+ ions are actively pumped out of the cell, maintaining a low concentration inside the cell ◆ Glucose passes from the inside of the cell into the capillaries by facilitated diffusion
Amino acids	epithelium; capillary; lumen of ileum; Na^+; amino acid; Na^+; ATP; ADP; amino acid; co-transport protein in plasma membrane of microvillus; amino acid transport protein if the sodium ion concentration in the epithelial cell becomes too high, the co-transport proteins will not take up as much amino acid	◆ Amino acids pass into the cell by diffusion ◆ There is a specific carrier molecule that takes both Na^+ ions and amino acid molecules into the cytoplasm together ◆ Na^+ ions are actively pumped out of the cell, maintaining a low concentration inside the cell ◆ The amino acids pass from the inside of the cell into the capillaries by facilitated diffusion

Fatty acids and glycerol	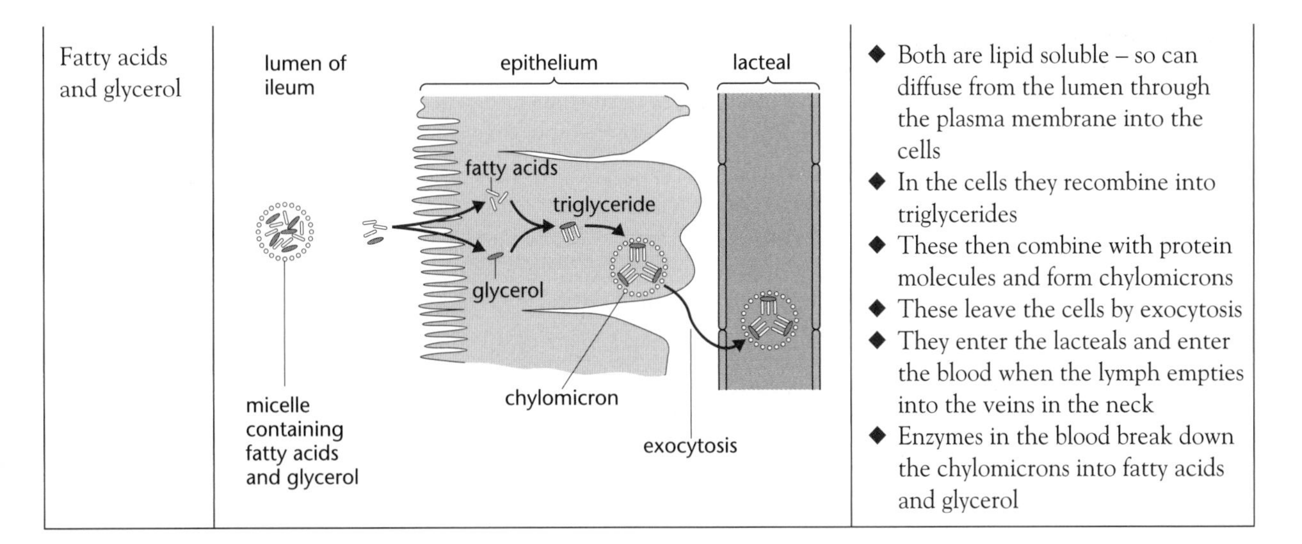	◆ Both are lipid soluble – so can diffuse from the lumen through the plasma membrane into the cells ◆ In the cells they recombine into triglycerides ◆ These then combine with protein molecules and form chylomicrons ◆ These leave the cells by exocytosis ◆ They enter the lacteals and enter the blood when the lymph empties into the veins in the neck ◆ Enzymes in the blood break down the chylomicrons into fatty acids and glycerol

Controlling digestion

It is energy-efficient to release enzymes only when they are required. The control of when they are released relies on a combination of **nervous control** and **hormonal control**.

Part of gut	Stimulus	Effect
Buccal cavity	◆ Smell or sight of food ◆ Food in contact with taste buds on the tongue	◆ Conditioned reflex leading to the secretion of saliva before food enters the buccal cavity ◆ Reflex leading to secretion of saliva onto the food
Stomach	◆ Food in contact with taste buds on the tongue ◆ Food in stomach	◆ Reflex leading to secretion of gastric juice ◆ Secretion of the hormone gastrin. This causes the release of gastric juice into the stomach
Pancreas and gall bladder	◆ Food in small intestine	◆ Secretion of the hormone secretin. This hormone causes the release of: – alkaline fluid by the pancreas ◆ Secretion of the hormone CCK-PZ (cholecystokinin-pancreozymin) ◆ This hormone causes the release of: – digestive enzymes by the pancreas – bile by the gall bladder

Diet and development in insects

- ◆ Many animals do not eat the same food throughout their lives.
- ◆ Some animals have two distinct phases in their life history:
  - – **larval stage** in which the animal grows but does not reproduce
  - – **adult stage** in which the animal does not grow but reproduces

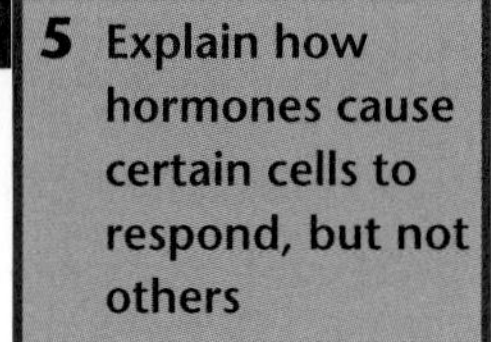

5 Explain how hormones cause certain cells to respond, but not others

- In some insects, the change in body form is a sudden process called **metamorphosis**.
- Examples of such a life history include lepidoptera, **butterflies and moths**.
- Both larval and adult butterflies have adaptations which allow them to exploit different sources of food.
  - Their mouthparts are very different. The larva has mandibles which cut and grind leaves on which it feeds. The adult has a proboscis with which it sucks nectar from flowers.
  - They produce different digestive enzymes as shown in the table below.

Stage in life cycle	Food	Digestive enzymes produced
Caterpillar (larva)	Leaves	Amylase, maltase, sucrase, protease and lipase
Butterfly (adult)	Nectar	Sucrase

- The enzymes produced by the caterpillar allow it to digest starch, protein and lipids found in the leaves, enabling growth.
- Sucrase hydrolyses sucrose, only found in nectar, to glucose and fructose. These monosaccharides can be respired to release energy for reproduction and dispersal.

WORKED EXAM QUESTIONS

1 a) Termites and peacock butterfly larvae are insects which require a large amount of protein in their diets. Both insects release enzymes from their gut tissues. These enzymes digest proteins to peptides and peptides to amino acids.

S (i) Name the chemical process responsible for the digestion of a peptide molecule into amino acid molecules. *(1 mark)*

Hydrolysis

> This question is synoptic, note the letter '*S*' in the margin, and relates back to module 1. When a question asks for a name, that is all you need to give. The answer looks a little bare and many candidates try to write full sentences but that is unnecessary. This word alone scores the mark.

(ii) Explain why a peacock butterfly larva requires a large amount of protein in its diet. *(1 mark)*

For growth

> Read the question very carefully, many candidates missed the word 'larva' and found it very hard to explain why the adult butterfly needed lots of protein. This candidate has scored the one available mark and has not wasted her time in expanding the answer. 'Growth' does explain why protein is needed by the larva.

b) The peacock butterfly larva feeds on leaves, a diet rich in cellulose and protein. Termites feed on wood, a diet rich in cellulose but poor in protein. Peacock butterfly larvae and termites benefit from the presence of mutualistic microorganisms which live in their guts.

The diagram summarises the biochemical processes carried out by these different microorganisms.

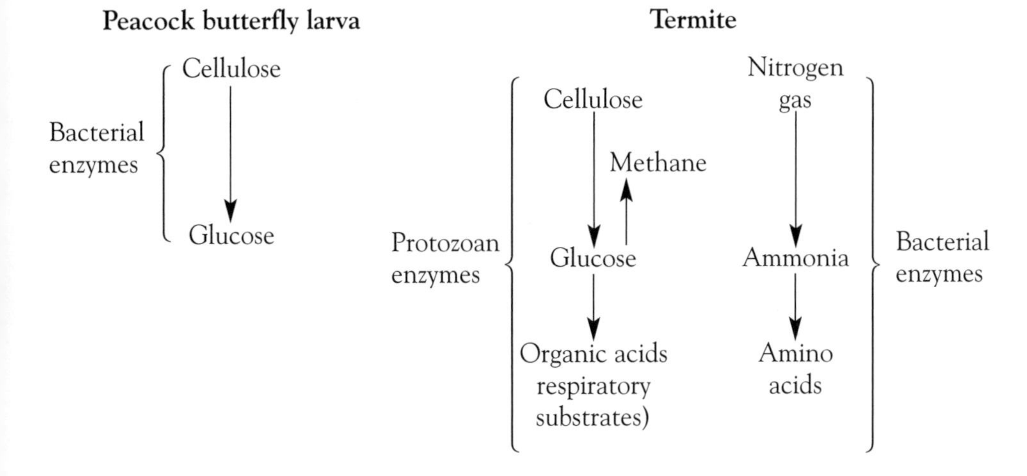

(i) Explain how it is possible for termites to survive on a diet poor in protein. *(2 marks)*

Protein is made up of amino acids. Bacteria in the gut of the termite produce amino acids. These molecules can be absorbed by the termite as if they had been produced by digestion of protein that it had eaten.

An excellent answer linking together a knowledge of the structure of protein and the data provided in the diagram above. However, maximum marks would have been given for the second sentence alone: there are bacteria/bacterial enzymes; which produce amino acids.

(ii) Peacock butterfly larvae obtain more energy from the same amount of cellulose than do termites. Explain why. *(1 mark)*

Bacteria in the larva produce glucose.

This answer does not really make the difference between the situations in the two organisms clear. Glucose is produced in the termite too, but in this case the glucose is used to produce methane or organic acids, reducing the amount available to the termite.

(iii) Nitrogen gas in the termite gut comes from swallowed air. However nitrogen fixation can only occur in anaerobic conditions, since the enzyme responsible for converting nitrogen gas to ammonia is inhibited by oxygen. Suggest how it is possible for nitrogen fixation to take place inside the termite gut. *(2 marks)*

The gut is made up of many active cells. These cells will absorb the oxygen from the air swallowed by the termite to use for respiration.

The question asks for a suggestion. That means you will not have been taught this, but will be able to work out the answer from your knowledge and the information provided in the stem of the question. The secret to answering 'suggest' questions successfully is 'DO NOT PANIC'.
The candidate gives one sensible idea, that oxygen may be used for aerobic respiration; it is more likely to be useful for the microorganisms in the gut, but the principle is acceptable. There are, however, 2 marks available, and she has missed the point that nitrogen makes up almost 80% of the air. Thus, the termite only needs a small amount of air to provide the nitrogen for the bacteria; and that amount of air will contain very little oxygen.

S c) The diagram shows part of an epithelial cell from a termite gut as seen under the electron microscope.

Give the letter and name of **two** features shown on the diagram and explain how each is involved in the rapid uptake of amino acids from the gut. *(4 marks)*

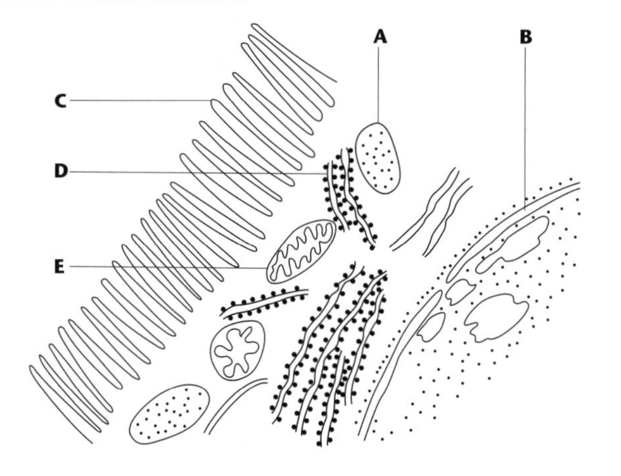

C
villus
Makes a greater the surface area for diffusion
E
mitochondria
produces energy for active uptake

This question was marked in two sections. Candidates that recognised and correctly named two labelled features scored one mark. A – lysosome; B – nuclear membrane C – microvillus; D – rough endoplasmic reticulum; E – mitochondrion. This candidate has made a typical error, and has confused the multicellular, finger-like projections of the gut, the villis, with the finger-like projections of a cell membrane, the microvillis. However, errors are never carried forward, and therefore the explanation of the involvement of structure C is correct. This is again a synoptic question, and the candidate ought to have used detailed knowledge from module 1 about of the role of the mitochondria. Remember they do not 'produce energy': they can 'release energy' or 'produce ATP'. So although the idea was correct, she lost the mark because of poor use of biological terminology.

d) In mammals, secretion of digestive juices is controlled by hormones and nerves.

(i) Describe **two** ways in which hormonal control differs from nervous control. *(2 marks)*

1. ***Hormone control is slow but it lasts longer***
2. ***Nervous control is fast and is done by impulses in nerves***

If asked for a comparison, always give both sides of the argument. This candidate has done that, but has offered two aspects of hormone control and then two aspects of nervous control. You can see that she has only given both sides of one difference; and therefore will only get one mark. To get the other mark, she would have to have written that hormonal control involves chemicals; or that nervous control only lasts for a short while. Do not be afraid to construct your own table, like the one given below:

Nervous control	Hormonal control
fast	slow
short term	long lasting

(ii) Secretin is a hormone. Explain its role in digestion. *(2 marks)*

Secretin is released into the bloodstream by the stomach. It circulates in the blood until it gets back to the stomach and is picked up by special receptor proteins on the surface of the lining cells. Those cells then release digestive juice into the stomach to break down food.

Even at A2 it is vital that you remember facts. There are three hormones related to digestion that are named in your specification, you must know what each of them does. This candidate has not only described the role of gastrin rather than secretin, but has also offered lots of unnecessary information on the way in which hormones function. The answer that was expected was that it causes the production/secretion of bile from the liver and alkali from the pancreas.

(AQA 2002)

EXAMINATION QUESTIONS

1 The table shows the diet of the large white butterfly and the enzymes it produces at different stages of its life cycle.

Stages in life cycle	Diet	Enzymes secreted by	
		salivary glands	mid-gut
Larva	Leaves	None	Amylase Maltase Proteases Lipase
Adult	Nectar	Sucrase	Sucrase

a) The larva and adult of the large white butterfly occupy different ecological niches. Explain how the diets and enzymes secreted are related to the demands of these stages of the life cycle. *(3 marks)*

S b) Leaves contain little glucose, yet large amounts of glucose can be found in the midgut of the larva of the large white butterfly. Explain why. *(2 marks)*

(AQA 2003)

13 Nervous systems and receptors

After revising this topic, you should be able to:

- understand that nervous communication involves detection of stimuli by receptors, transmission of nerve impulses by neurones and response by effectors
- describe the structure of a motor neurone
- explain how action potentials are initiated
- describe transmission across a synapse
- describe the structure of a Pacinian corpuscle
- explain how Pacinian corpuscles convert stimuli into electrical impulses in nerve cells
- describe the structure of a mammalian eye and its ability to focus images on the retina
- explain how rod cells and cone cells are used in monochromatic and trichromatic vision
- describe how rhodopsin causes a chemical change leading to the creation of a generator potential
- describe sensitivity and acuity of vision
- describe the pathway of a simple spinal reflex
- outline the functions of the parasympathetic and sympathetic nervous system
- explain taxes and kineses as simple responses which maintain an organism in a favourable environment

Important terms to understand

Term	Definition
Stimulus	◆ A change in the internal or external environment which can be detected in some way ◆ Stimuli are the signals to which an organism responds ◆ **External** stimuli may be features such as sound, pressure and light ◆ **Internal** stimuli may be features such as blood temperature
Receptor	A cell or an organ that is able to detect a particular stimulus which indicates a change in the environment of an organism. They have a number of characteristics. ◆ They are specific: a particular receptor will only respond to one kind of stimulus ◆ They are extremely sensitive and can detect small changes in the environment ◆ They convert the stimulus into an impulse
Response	An **action** in a muscle or a gland which arises as a result of stimulation by a nervous impulse.
Effector	A cell or an organ which responds to stimulation by the nervous system and produces a particular response. Muscles and glands are effectors, and respond to stimulation by either contraction or secretion.

Motor neurones

Feature	Description
Large cell body	This contains the nucleus and many of the other cell organelles.
Dendrons	These are short extensions of the cell body which transmit impulses **towards** the cell body.
Dendrites	◆ Each dendron has smaller extensions called dendrites. ◆ These dendrites are stimulated by other neurones and are points at which impulse transmissions always **start** in a motor neurone.
Axon	◆ It is a single extension of the cell body. ◆ It can be up to one metre long. ◆ It always transmits nerve impulses **away** from the cell body. ◆ It ends in a series of synaptic knobs, which are structures that stimulate the effector.
Myelin	◆ In mammals many axons are surrounded by a sheath of fatty material – **myelin**. ◆ The myelin enables the neurones to conduct nerve impulses rapidly.

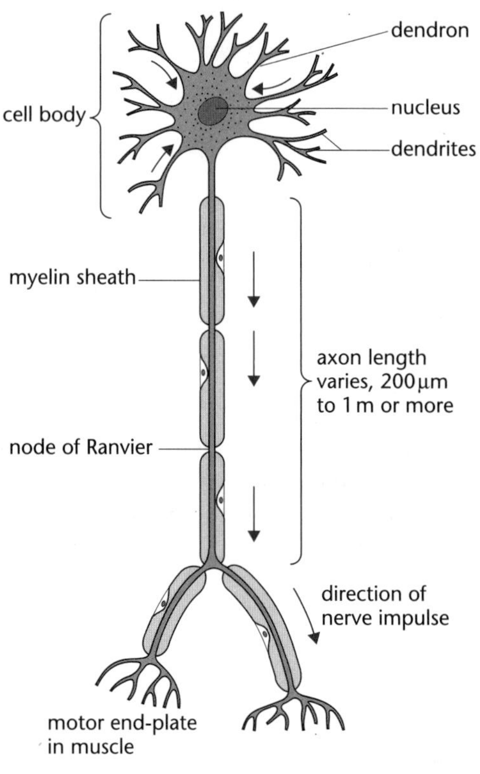

Figure 13.1 Structure of a motor neurone.

Nervous transmission

1 Name two other types of neurone.

A nerve impulse involves **changes in electrical potential** across the plasma membrane as the impulse passes along a neurone.

RESTING POTENTIAL

The resting potential is the potential difference (the difference in charge) across the membrane of a neurone when it is not conducting an impulse.

The distribution of ions inside and outside an axon of a typical mammalian neurone at rest is shown in the table below.

Ion	Concentration/mmol dm^{-3}	
	Inside axon	Outside axon
Potassium (K^+)	400	20
Sodium (Na^+)	50	460

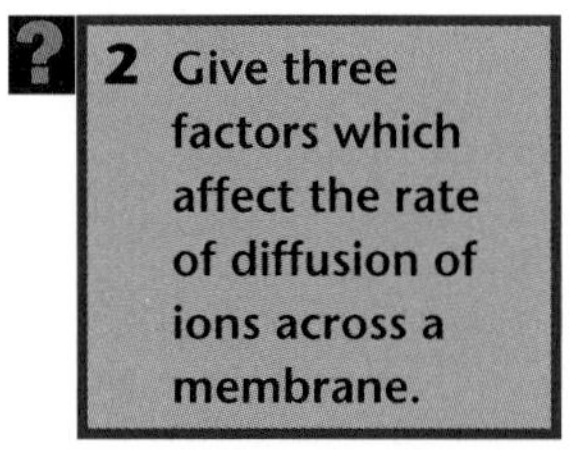
2 Give three factors which affect the rate of diffusion of ions across a membrane.

A resting axon is represented diagrammatically in Figure 13.2.

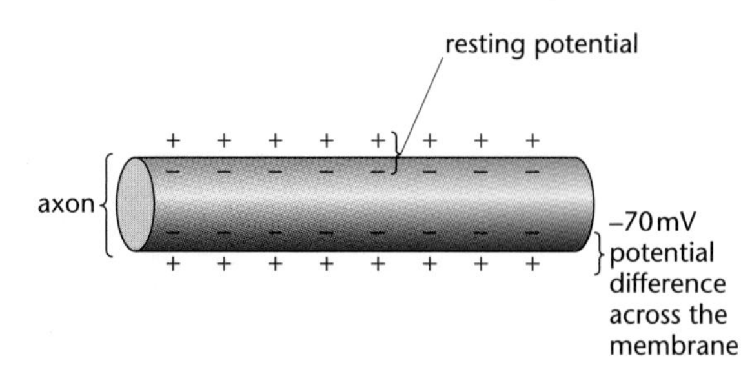

FIGURE 13.2 Resting potential in a neurone.

- There is an imbalance in the concentration of ions.
- There are more sodium ions on the outside of the cell.
- There are more potassium ions on the inside of the cell.
- These ions are able to diffuse through ion channels of the membrane of the neurone.
- This imbalance of ions is maintained by sodium-potassium pumps in the membrane of the neurone which use energy to remove sodium ions and take up potassium ions.
- The imbalance of ions results in a potential difference across the membrane.
- In mammals, this potential difference is about −70 mV.

EXAMINER'S TIP

Remember that it is sodium **ions** and potassium **ions** that move through the membrane of a neurone. It is not enough to just say sodium and potassium.

ACTION POTENTIAL

- The action potential begins when a particular stimulus causes the membrane at one part of the neurone to suddenly change its permeability.

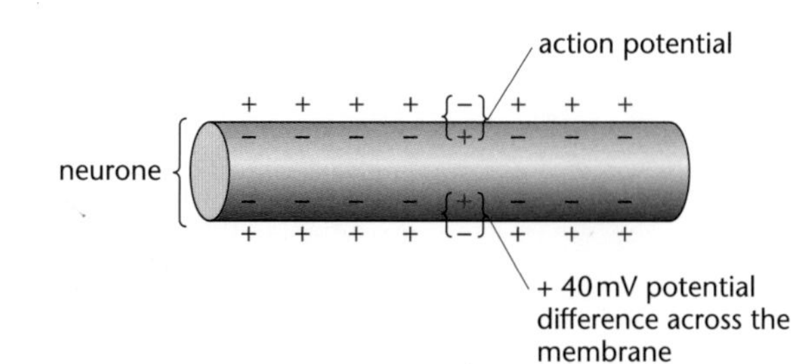

FIGURE 13.3 Action potential in a neurone.

- The inside becomes positive with respect to the **outside**.
- The membrane is said to have been **depolarised**.
- The resting potential is **re-established**.
- The membrane is said the be **repolarised**.

3 Suggest two types of stimulus that could initiate an action potential.

The changes in potential difference during an action potential are shown in Figure 13.4. These changes are due to changes in the permeability of the membrane to sodium ions and potassium ions.

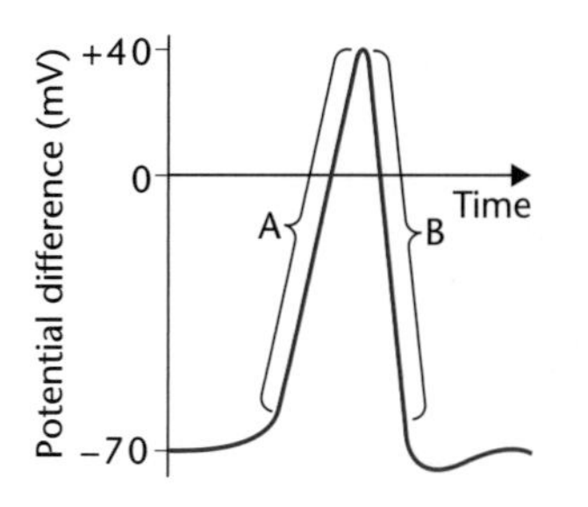

FIGURE 13.4 Action potential trace.

Part A (Depolarisation)	Part B (Repolarisation)
◆ The permeability to sodium ions increases. ◆ This is due to the opening of sodium gates. ◆ Sodium ions diffuse rapidly into the neurone. ◆ The potential difference across the membrane is +40 mV. ◆ The membrane is said to be depolarised.	◆ The sodium gates close. ◆ The permeability to potassium ions increases. ◆ This is due to the opening of potassium gates. ◆ Potassium ions flow out of the neurone. ◆ The potential difference returns to −70 mV. ◆ The membrane is repolarised.

EXAMINER'S TIP

Once the polarity across the membrane has changed during an action potential, it has to return to the resting potential before another action potential can occur. This period is called the **refractory period**.

THRESHOLDS AND ALL-OR-NOTHING

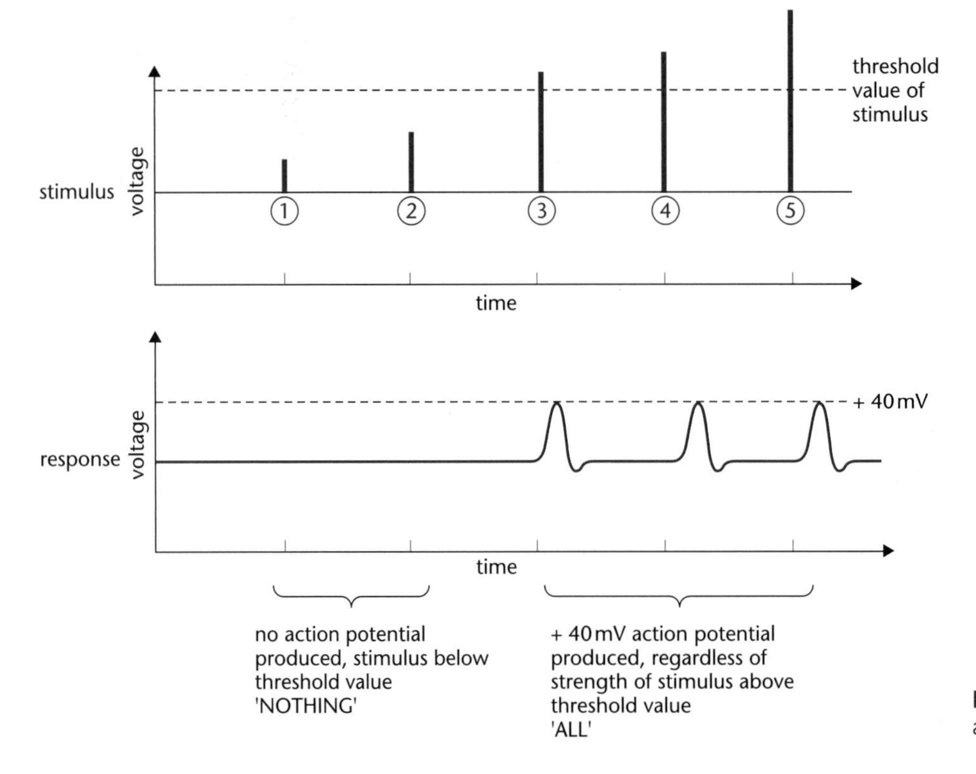

FIGURE 13.5 Thresholds and all-or-nothing.

Thresholds

- The **threshold value** is the lowest level of stimulus that will trigger an action potential.
- If the stimulus is too small, an action potential will not take place.

All-or-nothing

- Stimulus has to be at threshold level or above to cause an action potential.
- An action potential will **either be generated or not**.
- The resulting action potential is **always the same size**.
- You are aware of different strengths of stimulus due to the **frequency** of the action potentials. That is, how many impulses pass along a neurone in a given time.

FACTORS AFFECTING SPEED OF CONDUCTION OF IMPULSES

Saltatory conduction

- Some neurones are surrounded by a myelin sheath.
- Diffusion of sodium ions and potassium ions can only occur where there is no myelin sheath – at the **nodes of Ranvier**.
- An action potential can only occur at the nodes of Ranvier.
- Thus the impulse **jumps** from one node of Ranvier to the next.
- The speed of impulse transmission is up to 120 ms^{-1} – many times faster than non-myelinated neurones, which may be as low as 0.5 ms^{-1}.

Other factors

- **Temperature:** higher temperatures will increase the rate of diffusion of ions and thus increase the rate of conduction.
- **Diameter:** an axon with a larger diameter will transmit impulses faster.

The Synapse

Nervous pathways involve chains of at least two neurones that pass impulses from one to another. The junction between two neurones is called a **synapse**. The two neurones at a synapse do not touch each other. Instead, there is a gap. A simplified diagram of a synapse is shown in Figure 13.6.

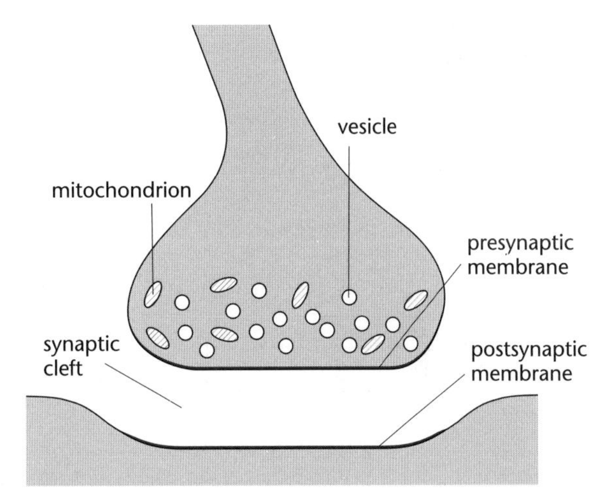

FIGURE 13.6 Structure of a synapse.

SYNAPTIC TRANSMISSION

Description	Diagrammatic representation
1. Nerve impulses arrive at presynaptic membrane.	nerve impulse ① presynaptic membrane synaptic knob
2. This stimulates calcium ion channels in the presynaptic membrane to open and allow calcium ions to diffuse into the neurone.	presynaptic membrane synaptic cleft ② calcium ions [Ca^{2+}]

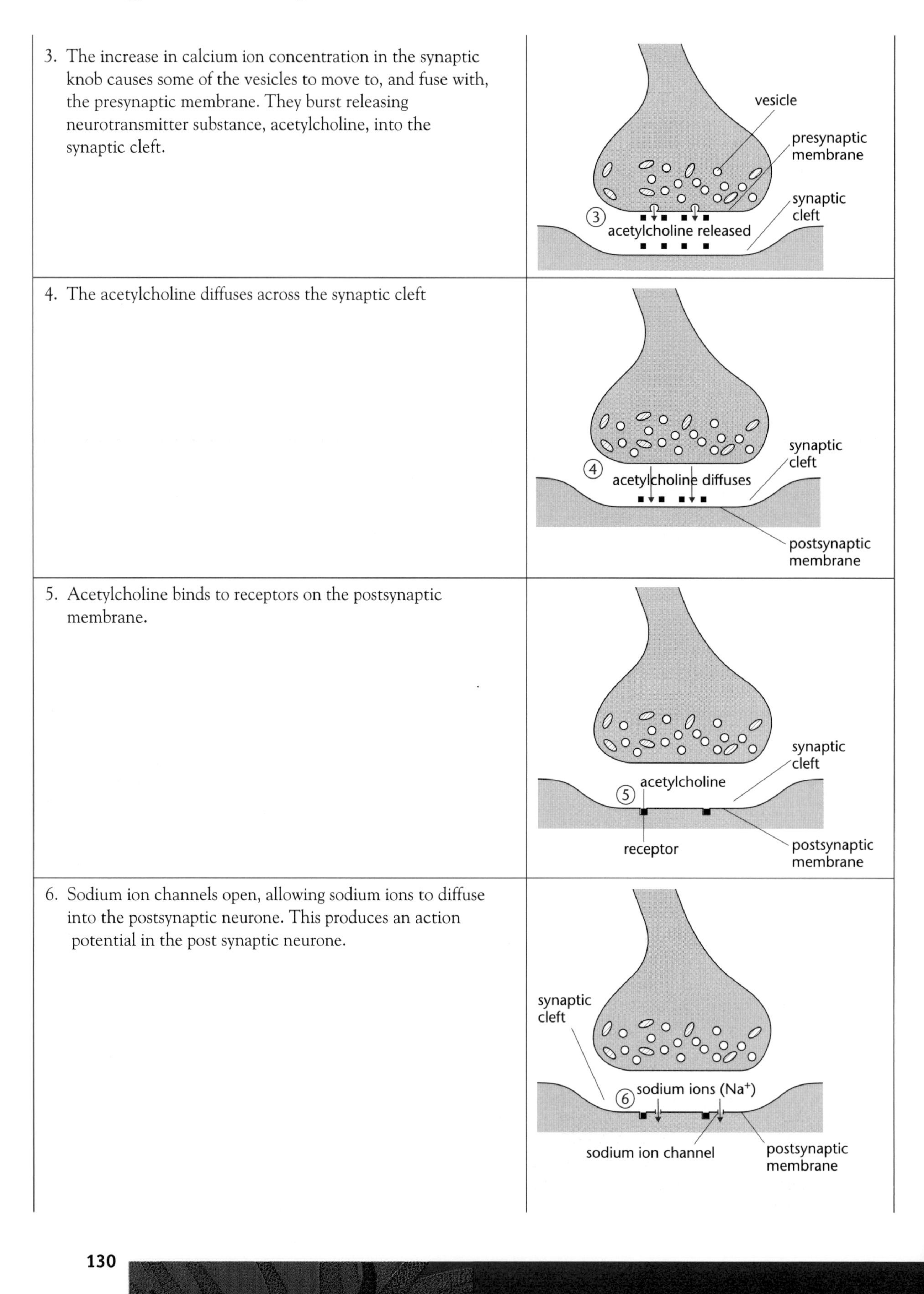
3. The increase in calcium ion concentration in the synaptic knob causes some of the vesicles to move to, and fuse with, the presynaptic membrane. They burst releasing neurotransmitter substance, acetylcholine, into the synaptic cleft.
vesicle
presynaptic membrane
synaptic cleft
3
acetylcholine released
4. The acetylcholine diffuses across the synaptic cleft
synaptic cleft
4
acetylcholine diffuses
postsynaptic membrane
5. Acetylcholine binds to receptors on the postsynaptic membrane.
synaptic cleft
acetylcholine
5
receptor
postsynaptic membrane
6. Sodium ion channels open, allowing sodium ions to diffuse into the postsynaptic neurone. This produces an action potential in the post synaptic neurone.
synaptic cleft
6
sodium ions (Na+)
sodium ion channel
postsynaptic membrane

7. The acetylcholine is hydrolysed into its components by the enzyme acetylcholinesterase. These components diffuse back across the synaptic cleft and are absorbed through the presynaptic membrane. They are used to produce more acetylcholine. Mitochondria in the synaptic knob provide ATP for resynthesis of the transmitter and for reforming vesicles.

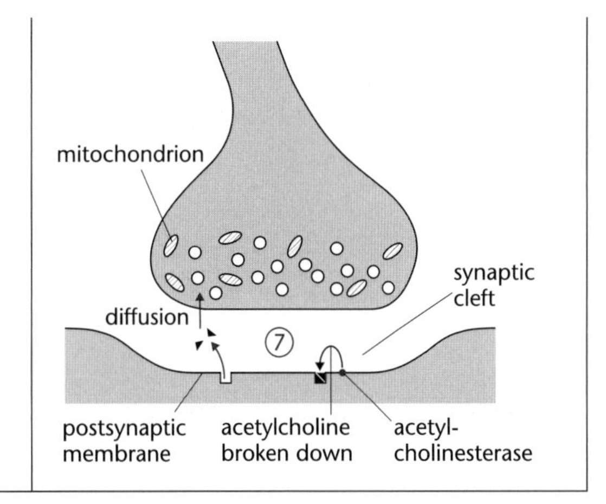

ASPECTS OF SYNAPTIC TRANSMISSION

Aspect	Explanation
Unidirectional	◆ Transmission at a synapse can only flow in **one direction**. ◆ Neurotransmitters are only released by the presynaptic membrane. ◆ Receptors for neurotransmitter are only present on the postsynaptic membrane.
Summation	◆ With a weak stimulus, not enough neurotransmitter will be released to reach a threshold level. ◆ Therefore depolarisation in the postsynaptic membrane does not occur. ◆ Weak stimuli from many neurones cause **simultaneous release** of many molecules of the neurotransmitter. ◆ This results in depolarisation in the postsynaptic membrane.
Inhibition	◆ Different neurotransmitters may be released by the presynaptic membrane. ◆ They affect different receptors on the postsynaptic membrane. ◆ The resting potential **falls** to a lower level. ◆ The postsynaptic membrane is less likely to reach the threshold and an action potential is less likely.

Receptors

Receptors convert some form of energy into a **generator potential**, which may then initiate an **action potential** in a neurone. The action potential then passes along the neurone and is called an **impulse**.

? 4 Suggest how drugs could stop synaptic transmission.

There are many different receptors. The only ones you are required to know about are Pacinian corpuscles, and the rod and cone cells in the retina of the eye.

PACINIAN CORPUSCLES

Pacinian corpuscles are **pressure receptors** found mainly in the skin. They are also found in some joints and tendons. Figure 13.7 shows a diagram of a Pacinian corpuscle.

? 5 a) Give three different forms of energy that may be converted into an action potential.
b) Name the receptors that are specific to each form of energy you have named.

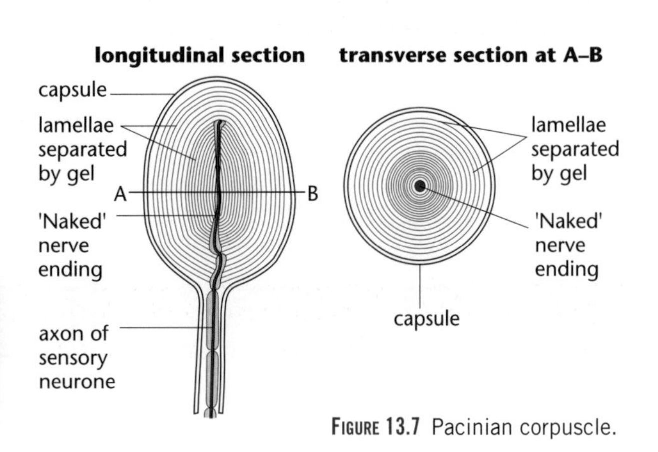

FIGURE 13.7 Pacinian corpuscle.

Event	Diagram representation
◆ Pressure on the skin is transmitted to the corpuscle ◆ If the pressure is great enough, it deforms the corpuscle	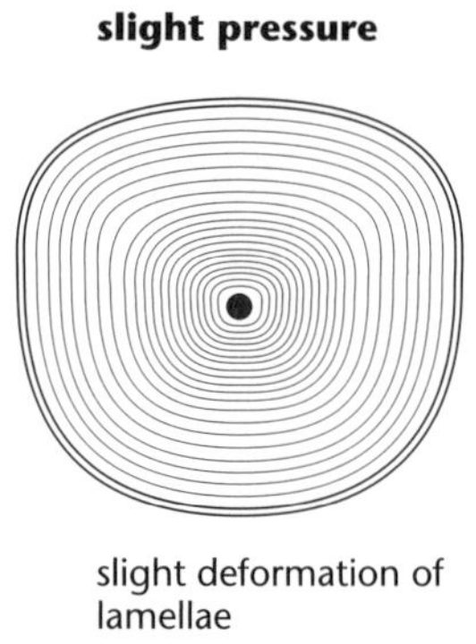 **slight pressure** slight deformation of lamellae
◆ This deformity will open pressure-sensitive sodium ion channels in the membrane of the corpuscle ◆ Sodium ions move inwards	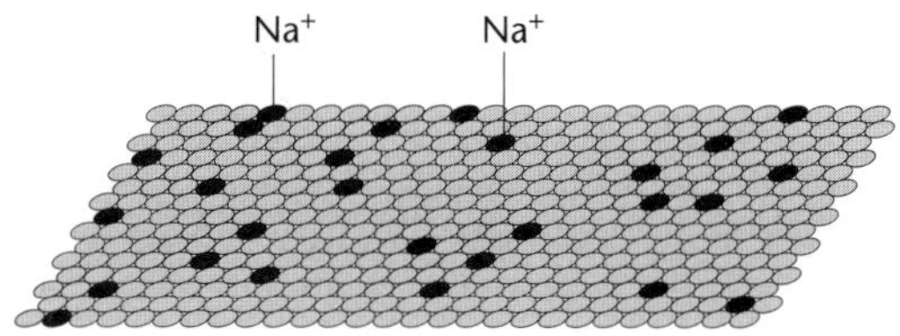 Na^+ Na^+ a few sodium ion channels in membrane are opened
◆ This changes the balance of charge across the membrane ◆ This change in membrane potential is called the generator potential	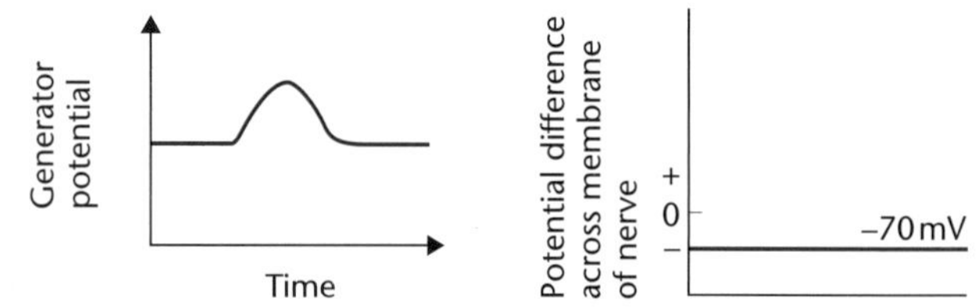 Generator potential; Time Potential difference across membrane of nerve; + 0 –; –70 mV
◆ A greater pressure will open more channels, producing a larger generator potential	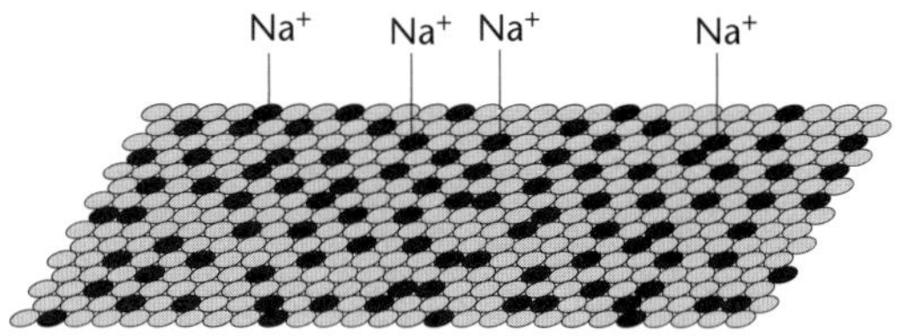 Na^+ Na^+ Na^+ Na^+ many sodium ion channels in membrane are opened
◆ If the generator potential is large enough, a nerve impulse will be initiated	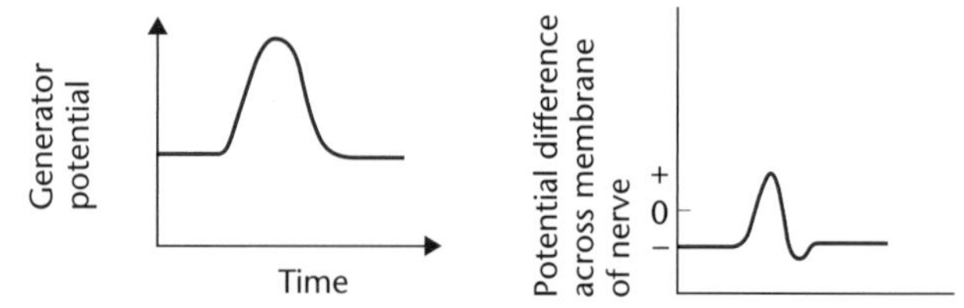 Generator potential; Time Potential difference across membrane of nerve; + 0 –

6 What other receptors are found in the skin?

The eye

STRUCTURE

You must be able to identify structures in the eye and know their function.

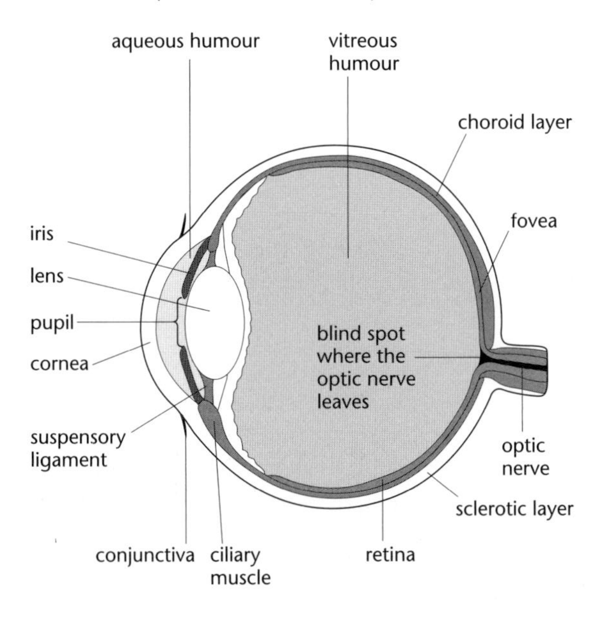

FIGURE 13.8 Structure of the eye.

Part of eye	Description	Function
Conjunctiva	Very thin transparent membrane covering the cornea	◆ Transmits light ◆ Protects cornea from friction
Cornea	Transparent front part of the sclerotic layer	◆ Transmits light ◆ Refracts (bends) light
Sclerotic layer	Collagen fibres making up the outer layer of the eye	◆ Protects the sensitive cells inside the eye ◆ Maintains the shape of the eyeball
Choroid layer	Black middle layer of wall of eye	◆ Dark colour prevents internal reflection of light rays ◆ Contains blood vessels
Iris	Coloured part of choroid layer	◆ Controls the amount of light entering the eye by changing the size of the pupil
Lens	Transparent crystalline structure	◆ Transmits light ◆ Refracts light
Ciliary muscles	Ring of muscle outside lens	◆ Changes the shape of the lens and therefore its ability to refract light ◆ Muscles contract to produce a more convex lens with greater refracting power ◆ Muscles relax to produce a less convex lens with less refracting power
Retina	Inner layer of wall of eye containing rod cells and cone cells	◆ Rod and cone cells convert light energy into generator potentials and initiate a nerve impulse ◆ Fovea has the greatest concentration of cone cells and no rod cells ◆ Blind spot has no rod or cone cells
Aqueous and vitreous humours	Fluids in the eye: aqueous humour is less viscous than vitreous humour	◆ Transmit light ◆ Maintain shape of eyeball

EXAMINER'S TIP

Light is **refracted** (bent) as it moves **from one medium to another**. Therefore, the cornea, aqueous humour and lens are all involved in accommodation.

ACCOMMODATION

Light from a point on an object spreads out. Any of the rays of light from that point that enter the eye are bent so that they hit a single point on the retina. This is called focusing.

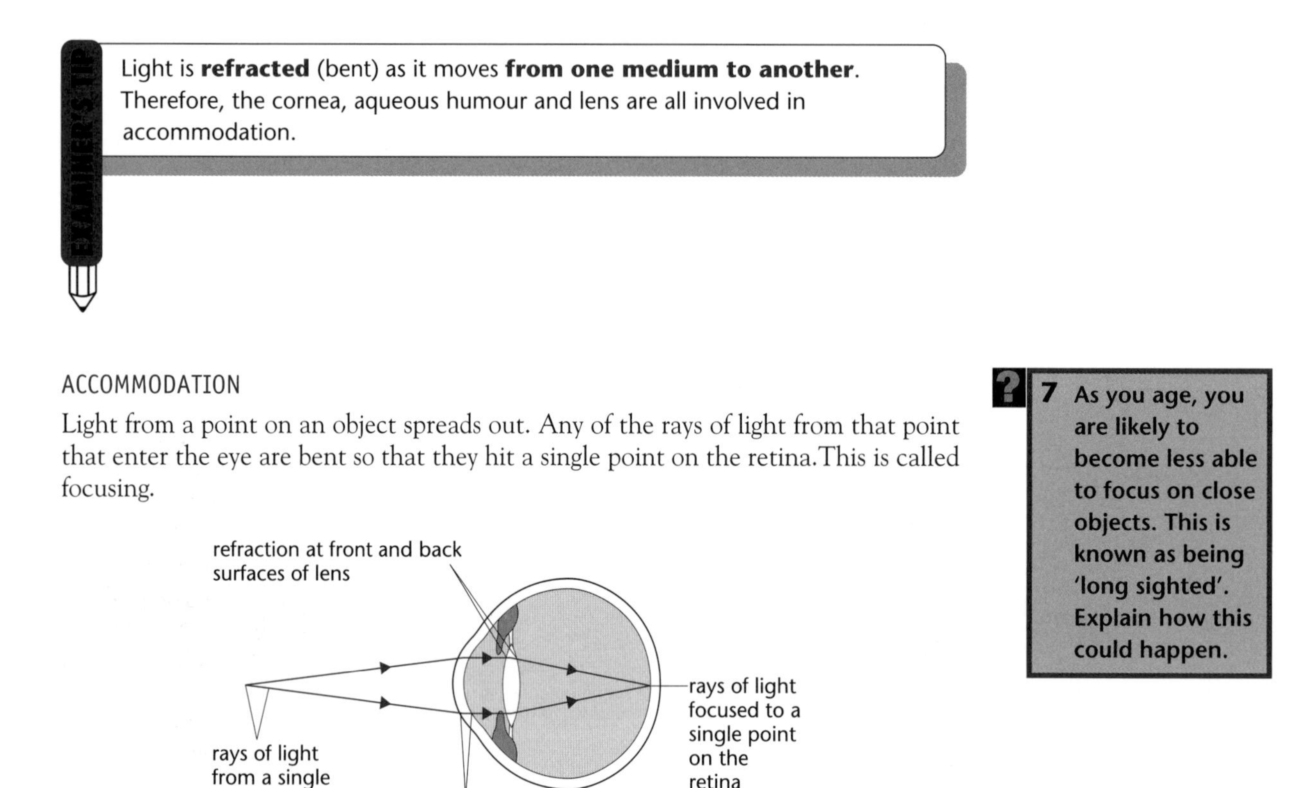

FIGURE 13.11 How the eye focuses light.

7 As you age, you are likely to become less able to focus on close objects. This is known as being 'long sighted'. Explain how this could happen.

The ability of the eye to focus on near and then on distant objects is called **accommodation**.

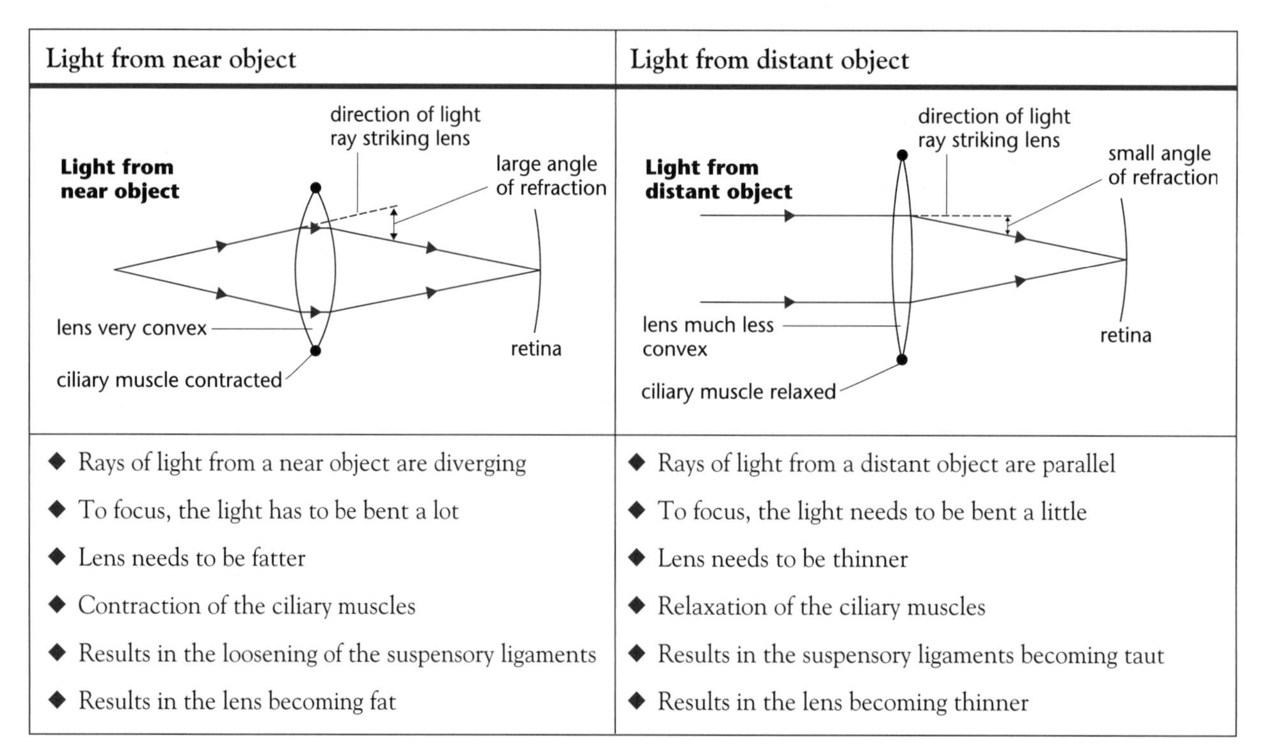

Light from near object	Light from distant object
◆ Rays of light from a near object are diverging	◆ Rays of light from a distant object are parallel
◆ To focus, the light has to be bent a lot	◆ To focus, the light needs to be bent a little
◆ Lens needs to be fatter	◆ Lens needs to be thinner
◆ Contraction of the ciliary muscles	◆ Relaxation of the ciliary muscles
◆ Results in the loosening of the suspensory ligaments	◆ Results in the suspensory ligaments becoming taut
◆ Results in the lens becoming fat	◆ Results in the lens becoming thinner

ROD AND CONE CELLS

These are the **light-sensitive receptors** found in the retina.

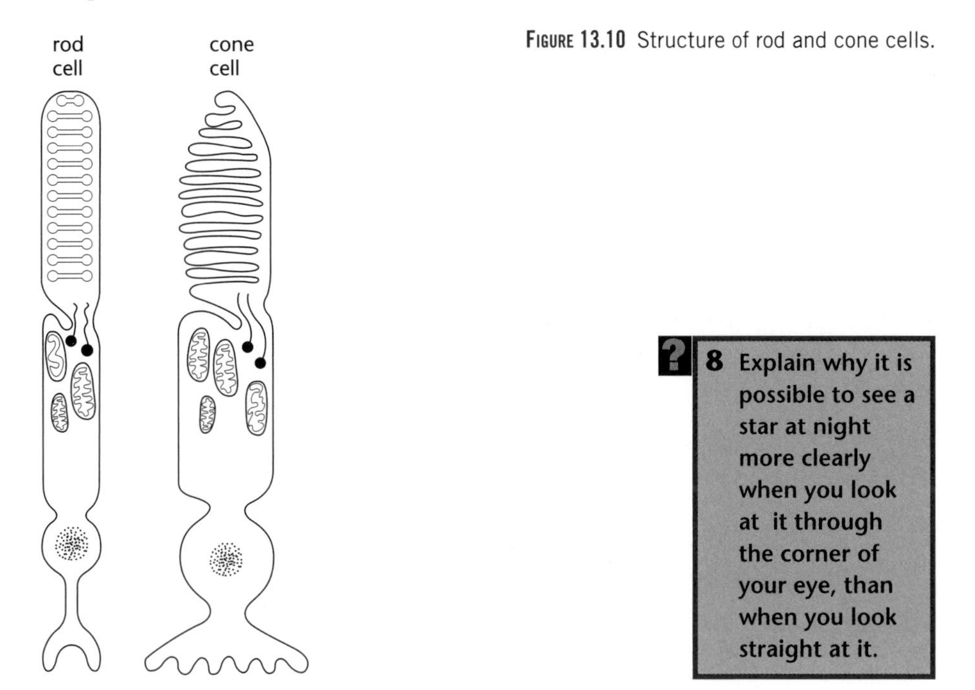

FIGURE 13.10 Structure of rod and cone cells.

8 Explain why it is possible to see a star at night more clearly when you look at it through the corner of your eye, than when you look straight at it.

Feature	Rod cells	Cone cells
Distribution	◆ Evenly throughout the retina ◆ Absent from the fovea	◆ Present mainly in the fovea
Sensitivity to light	◆ Very sensitive to light, therefore operate in dim conditions. This is because neurotransmitter released from several rod cells can combine, and so the threshold is more likely to be reached and an action potential initiated. This is called **summation** and is possible because several rods are linked to one neurone. This is known as **retinal convergence**. ◆ Insensitive to colour.	◆ Less sensitive to light, therefore only operate in bright light. Single cone cell linked to one neurone. Therefore each cone cell must produce enough neurotransmitter to reach the threshold needed for an action potential. In low light intensities this is unlikely. ◆ Three types of cone cell: one sensitive to red light, one sensitive to green light, and one sensitive to blue light.
Visual acuity	◆ Produces poorly resolved (unclear) images. This is because several rods are connected to the same bipolar cell.	◆ Produces well-resolved (clear) images. This is because each cone cell is connected to one bipolar cell.
Light sensitive pigments	◆ Single pigment, called rhodopsin, in every rod cell.	◆ One of three types of pigment in each cone cell. ◆ The pigments are sensitive to red, green or blue light. ◆ Stimulation of different combinations of the three types of cone cells produces coloured vision.

CREATING A GENERATOR POTENTIAL

When sense cells in the eye are stimulated by light, a change occurs in a photosensitive pigment. The pigment in rod cells is **rhodopsin** and the changes that occur on stimulation are shown in Figure 13.11.

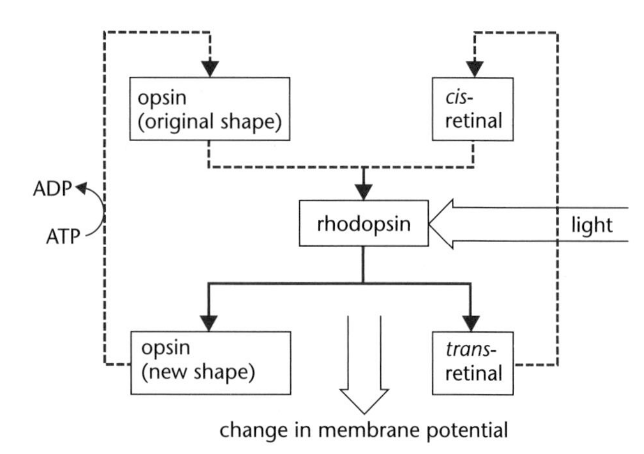

FIGURE 13.11 Change in membrane potential.

The spinal cord and spinal reflexes

- A simple reflex is inborn (not learned), and always results in the **same fixed response** to a particular stimulus. Examples of reflexes include dilating the pupil in dim light, salivating at the taste of food and the knee jerk reflex.
- The nervous pathway of a simple reflex is called a **reflex arc**.

The general layout of a three-neurone reflex arc is shown in Figure 13.12.

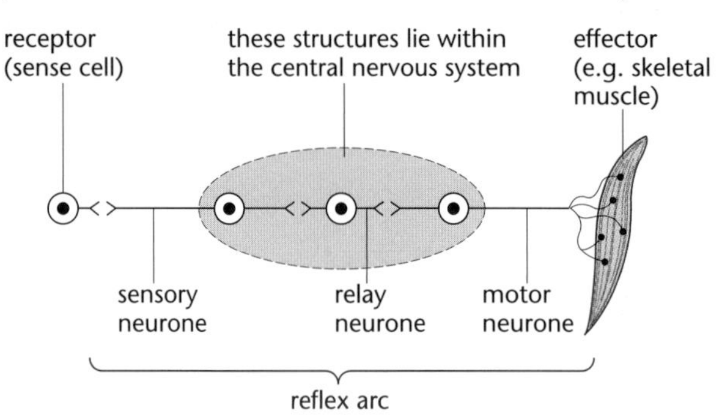

FIGURE 13.12 Simple reflex arc.

Reflex arcs that involve the spinal cord, but not the brain, are called spinal reflexes.

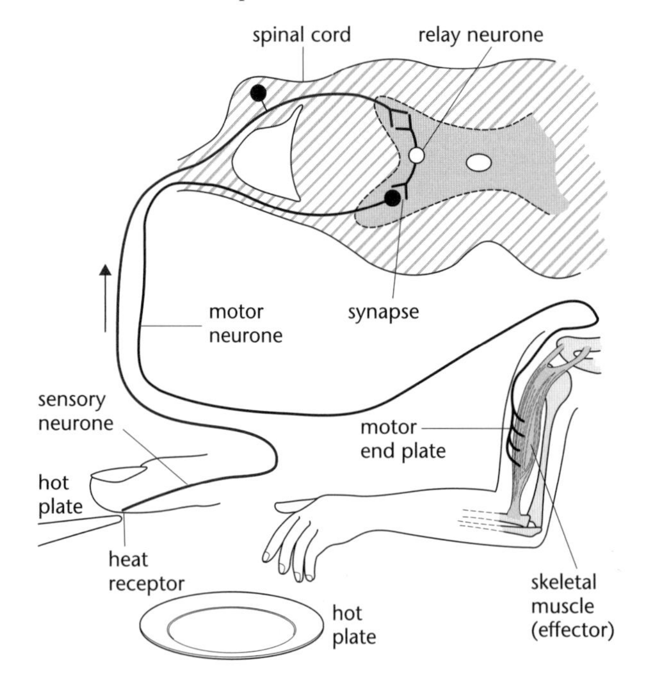

FIGURE 13.13 Specific reflex arc.

ROUTE OF IMPULSE

- Your hand touches a very hot plate.
- The heat receptors and pain receptors in your skin detect the stimulus.
- They produce generator potentials and initiate action potentials in the sensory neurone.
- The sensory neurone carries impulses into the spinal cord via the dorsal root.
- In the grey matter of the spinal cord
  - the **sensory** neurone synapses with a **relay** neurone
  - the **relay** neurone synapses with **motor** neurone
- The motor neurone carries impulses out of the spinal cord via the ventral root.
- The impulses are carried to the muscles.
- The motor neurone releases a neurotransmitter substance onto muscle cells causing them to contract.
- Your hand is moved away from the very hot plate.

Autonomic nervous system

This is a part of the nervous system that controls internal glands and muscles, which are **not under conscious control**.

It has two sections:

1. The SYMPATHETIC NERVOUS SYSTEM

- The motor neurones release the neurotransmitter **noradrenaline** onto the glands or organs they stimulate.
- It plays an important part in enabling the body to react to stress.
- An example of an effect is increasing heart rate.

2. The PARASYMPATHETIC NERVOUS SYSTEM

- The motor neurones release the neurotransmitter **acetylcholine** onto the glands and organs they control.
- This has an inhibitory effect, and is important when the body is at rest.
- An example of an effect is decreasing heart rate.

Simple behaviour in animals

The behaviour of some animals consists almost entirely of simple reflexes. Two types of simple behaviour, **kinesis** and **taxis**, help maintain animals in a favourable environment.

TAXIS

- This is a **directional** response to a **stimulus**.
- Organisms will either move towards or away from a directional stimulus.
- For example, maggots move away from light.
- This will move them away from an area where they are more likely to be preyed on.

KINESIS

- This is a **non-directional** response to a **stimulus**.
- Organisms will either increase or decrease their movement as the intensity of the stimulus changes.
- For example, woodlice move less in moist conditions and more in dry conditions.
- Woodlice lose water easily and dehydrate in dry conditions. By moving less in moist conditions, they are likely to remain in these favourable conditions. Conversely, if they move more in dry conditions, they are likely to leave the dry conditions and find more favourable, moist conditions

WORKED EXAM QUESTIONS

1 **Figure 1** shows two motor neurones, **A** and **B**. It also shows the synapses of neurone **B** with three other neurones **P**, **Q** and **R**.

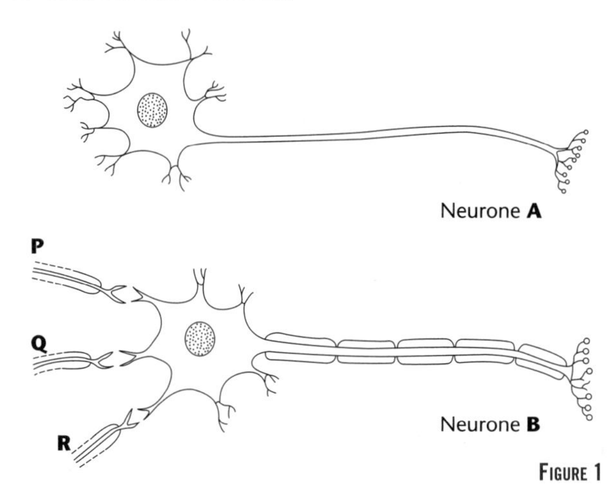

FIGURE 1

a) (i) An action potential is produced in neurone **A**. Describe how this action potential passes along the neurone. *(3 marks)*

When an action potential is produced sodium enters the neurone and the outside of the neurone becomes negatively charged when compared with the inside. We call this change depolarised because the membrane was polarised before. Movement of molecules like sodium annoys neighbouring parts of the membrane and they depolarise too. This is how the action potential moves along the neurone.

This question is not about the action potential but how it passes along the neurone. This candidate has focused on the key phrase 'action potential' and has spent too long explaining that term. He has also forgotten the first rule in terms of action potentials, that is, that **ions** move through the membrane. To write 'sodium' is inaccurate; to write 'molecules of sodium' is wrong. However, he does have the idea that the adjoining region of the membrane depolarises and will get credit for this. He has missed the points that local currents are created by an action potential, which change the permeability of the adjoining membrane, allowing sodium ions to enter. Try to avoid using emotional ideas like 'annoyed' – the membrane does not get annoyed.

(ii) Explain why the transmission of a series of nerve impulses along neurone **B** uses less energy than transmission along neurone **A**. *(3 marks)*

Neurone B has a sheath around it made of myelin and an action potential can only be produced between sections of the sheath. These areas are called nodes of Ratner. The impulse therefore jumps from one node to the next rather than to bumble along the whole of the neurone. This is much quicker and uses less energy.

He has clearly made the distinction between the structures of the myelinated and non-myelinated neurones and has described saltatory conduction for two marks. However, he has not explained why that should mean that one neurone uses less energy than the other. Restoring the resting potential; an energetic process; involving the active transport of sodium ions out of the membrane; only occurs at the nodes of Ranvier (not Ratner – it is obvious what the candidate means, but be careful because slips like this could cost marks) in myelinated neurones, whereas in non-myelinated neurones this process takes place all the way along.

Teachers try very hard to give students a picture of complicated processes and we can all appreciate the action potential 'bumbling along' the non-myelinated neurone, but try to avoid using these phrases in your exam answers.

b) Neurones of type **A** are found in the autonomic nervous system. The autonomic nervous system consists of the sympathetic division and the parasympathetic division.

(i) Suggest the effect that stimulation by neurones of the sympathetic division would have on the diameter of arterioles leading to skeletal muscle. Explain your answer. *(2 marks)*

The sympathetic division of the autonomic nervous system prepares the body for activity. So if the muscle is to be more active it needs food and oxygen to provide energy. That comes from the blood, so the artery will get wider.

The candidate worked out the answer that the arteriole (not the artery – a lack of care on his part) would dilate. However, the arteriole always provides blood to the muscle. When it is dilated, it provides **more**, and that idea was worth the second mark.
There is no point in explaining the role of the sympathetic division of the autonomic nervous system unless you are asked for it.
Try to avoid imprecise terms like 'food' when you mean organic molecules.

S (ii) Explain the effect of the parasympathetic division of the autonomic nervous system on cardiac output. *(4 marks)*

The parasympathetic division of the autonomic nervous system prepares the body for inactivity. So it reduces cardiac output. In this way there is less blood passing around the body carrying less food and oxygen so the cells will not be able to be so active.

Watch for the **'S'** in the margin, it is a clue that the answer to this question should include some material from other modules: it stands for the word **synoptic**.
Control of cardiac output is in Module 1, and so in this answer the candidate should go into detail of the heart and how the nervous system alters its activity.
This candidate would not get any marks for this answer as he did not mention the SA node; the vagus nerve that affects it; how it is affected (its activity decreases) or that this decreases the heart rate.

c) (i) **Figure 2** shows the effect of impulses from neurones **P** and **Q** on the production of an action potential in neurone **B**.

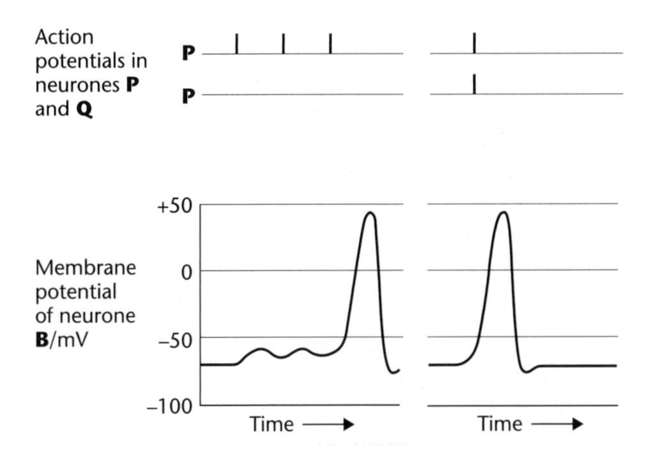

FIGURE 2

Each effect is a type of summation. Use information in **Figure 2** to explain the two types. *(2 marks)*

First type. Many impulses from neurone P arrive close together and cause an impulse in neurone B.

Second type. An impulse from neurone P and one from neurone Q arrive at the same time and cause an impulse in neurone B

It is important to use all the information given in the question. In this case, the neurones P, Q and R are shown making synaptic links with neurone B in Figure 1. By this time, most candidates had forgotten that, which made this question more difficult to answer.
This candidate has done exactly what the question asked him to do. He has used the information in the diagram to explain the difference between the two types of summation. The first is temporal summation, the second spatial summation, but those names were not needed for full marks.

(ii) **Figure 3** shows the effect of an impulse from neurone **R** on the membrane potential of neurone **B**.

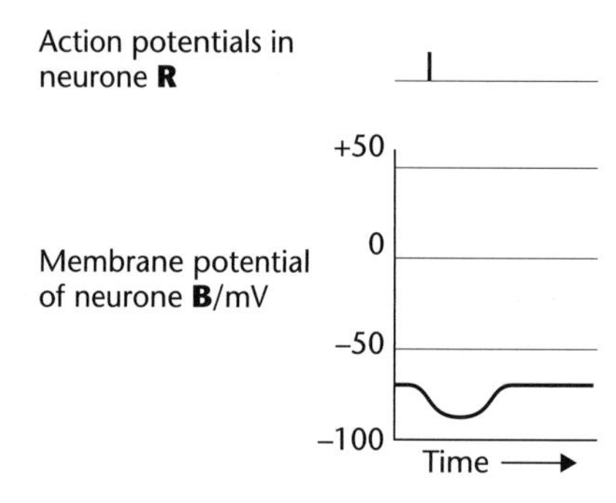

FIGURE 3

Describe the kind of synapse between neurone **R** and neurone **B**. *(2 marks)*

As the impulse from neurone R causes the resting potential to get lower this must be an inhibitory synapse.

No need to explain the answer, although this candidate did try.
Describe the synapse … means … what is the synapse?
It is an **inhibitory** synapse. That answer was all that was expected.

(AQA 2003)

EXAMINATION QUESTIONS

1 **Figure 1** shows a Pacinian corpuscle and its sensory neurone which are present in the skin of a fingertip.

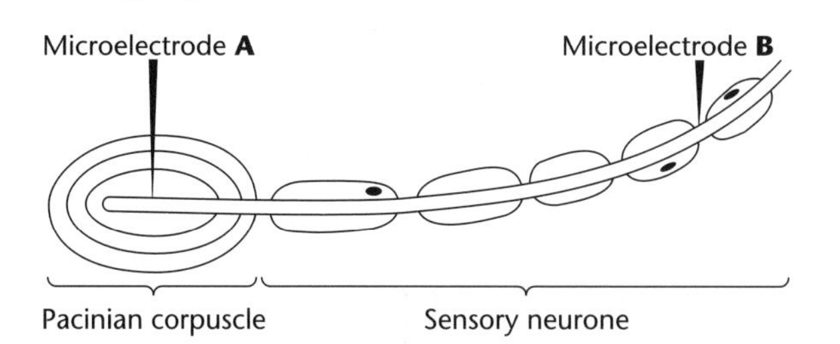

FIGURE 1

Figure 2 shows the electrical activity simultaneously recorded from the Pacinian corpuscle and its sensory neurone when increasing pressure was applied to a fingertip.

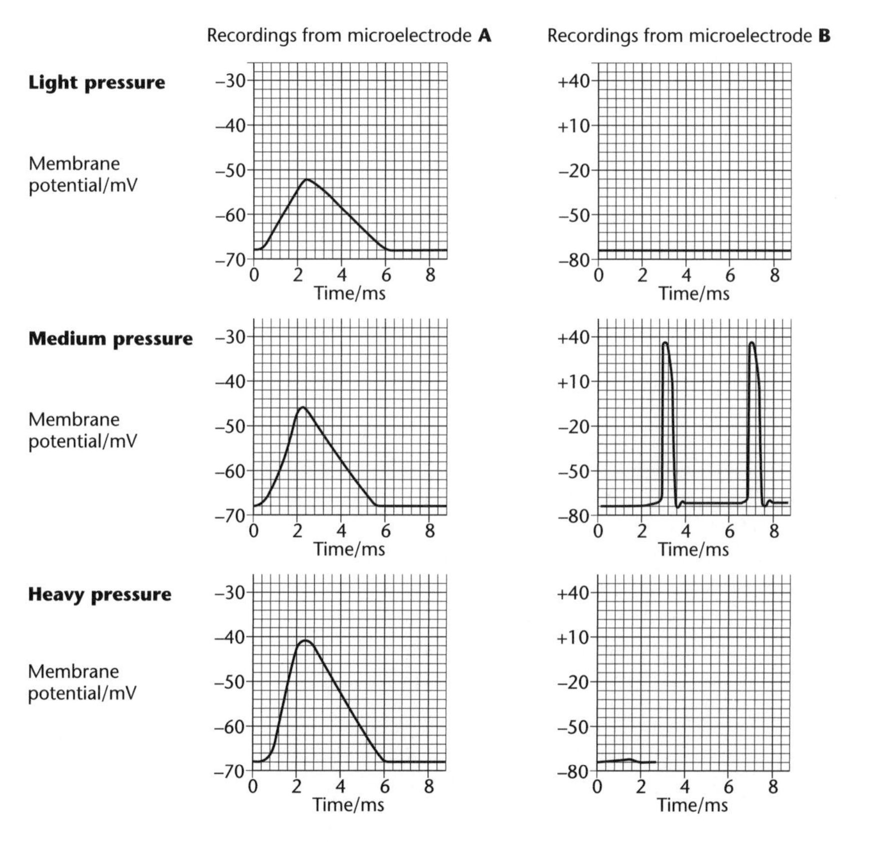

FIGURE 2

a) Explain how pressure on the Pacinian corpuscle produces the changes in membrane potential recorded by microelectrode **A**. *(2 marks)*

b) (i) Draw an arrow on **Figure 1** to show the direction of net movement of potassium ions during repolarisation of the sensory neurone. Label this arrow with the letter **K**. *(1 mark)*

(ii) Complete **Figure 2** to show the expected electrical activity recorded by microelectrode **B** when high pressure was applied to the fingertip. *(1 mark)*

(AQA 2002)

c) (i) What is the delay between the maximum depolarisation recorded by microelectrode **A** and the first depolarisation recorded by microelectrode **B** when medium pressure was applied to the fingertip? *(1 mark)*

(ii) The distance between microelectrodes **A** and **B** is 8 cm. Use this information together with your answer to c) (i) to calculate the speed of conductance of an impulse along the sensory neurone, in metres per second. Show your working. *(1 mark)*

d) Most of the sensory neurone in **Figure 1** is covered by a myelin sheath. This prevents the movement of ions across the axon membranes except at the small gaps in the sheath, called the nodes of Ranvier. Multiple sclerosis is a disease in which the myelin sheaths surrounding the neurones are destroyed so the neurones become de-myelinated.

(i) Explain how de-myelination of neurones produces slow responses to stimuli in people with multiple sclerosis. *(2 marks)*

(ii) The rate of ATP consumption of a de-myelinated neurone is greater than that of a myelinated neurone when conducting impulses at the same frequency. Explain why. *(2 marks)*

ANSWERS TO IN-TEXT QUESTIONS

CHAPTER 1

1 Interphase – which is divided into G1, S and G2 stages. Replication occurs during the S stage.

2 Nucleotides

3 Something along the lines of:
- a) big teeth
- b) pleasant smell; bright colours
- c) this depends on the person … but an example could include 'blue eyes'

4
- a) Red (R) – white (r)
- b) Smooth (S) – wrinkled (s)
- c) Green (G) – yellow (g)

5 Both parents with blood group O must be homozygous recessive for the O allele. Therefore the only allele in both gametes will be O. The child is 100% sure to be blood group O. Thus the probability will be 1.
There is a 1 in 2 chance of it being a girl.
Multiply those together: $1 \times \frac{1}{2} = \frac{1}{2}$ a 50% chance.

6
- a) If she inherits the recessive allele on both X chromosomes from her parents.
- b) Father must be X^h Y. Mother could be either $X^h X^h$ or $X^h X^H$

CHAPTER 2

1 The genotype determines which enzymes are made, which determines the measurable characteristics.

2 Random assortment of unlinked genes produces gametes with different genotypes; crossing over at chiasmata produces new combinations of alleles of genes on a chromosome.

3 The total number of alleles of a particular gene in the genotypes of individuals in a population.

4
- a) New individuals enter the population carrying a new ratio of the alleles of the gene in question.
- b) A change in one gene changes the existing ratio of alleles.

5 0.4 (1 – 0.6)

6
- a) The frequency of **Aa** is 0.48 ($2pq = 2 \times 0.4 \times 0.6 = 0.48$).
- b) The number of **Aa** individuals is 480 (1000×0.48).

7 Fitness is a measure of the fertility of an organism in a particular environment. An organism that is well adapted to its environment and is able to produce large numbers of offspring is said to be fit.

8 A gene mutation in one insect confers resistance to pesticide; insect with resistance gene less affected by pesticide; resistant insect produces more offspring than susceptible insects; resistant insect passes resistance gene on to offspring; frequency of resistance gene increases.

9
- a) Selection against individuals carrying the $\mathbf{Hb^S}$ allele, i.e. individuals carrying the $\mathbf{Hb^S}$ allele will have fewer offspring (less energy for successful pregnancy or offspring die in infancy). As a result the $\mathbf{Hb^S}$ frequency of the allele will decrease.
- b) Selection now against individuals who are $\mathbf{Hb^A\ Hb^A}$, i.e. such individuals will have fewer offspring (parents or children die of malaria). As a result, the frequency of the $\mathbf{Hb^A}$ allele will decrease.

10 Since the values of mean $\pm 1.96 \times$ SE did not overlap, we can be confident that the differences in the mean are not due to chance. The probability of them being due to chance is less than 5% ($p < 0.05$) so the differences are statistically significant.

CHAPTER 3

1 The chromosomes in the parental sets **A** and **T** are not homologous. As a result, meiosis cannot proceed. In the tetraploid **AATT**, homologous chromosomes occur in the two **A** sets of chromosomes and in the two **T** sets of chromosomes, so meiosis can proceed.

2
- a) Hybridisation and polyploidy
- b) Since $4n = 52$, $2n = 26$

3 Despite their prolonged reproductive isolation, they did not accumulate sufficient genetic differences to prevent their ability to interbreed and produce fertile offspring.

4 The desert prevents contact between the two populations of mice. This reproductive isolation is essential to speciation. Being active flyers, the birds of prey can cross the desert; so there is no reproductive isolation between the two populations.

5 *Homo*

6 Any two of the following pairs of contrasting statements:

Prokaryotic cells	Eukaryotic cells
◆ Smaller (0.5 to 5.0 μm in diameter)	◆ Larger (up to 40 μm in diameter)
◆ DNA is single-stranded, circular and not covered by proteins	◆ DNA is double stranded, covered by proteins (histones) and arranged in one or more linear chromosomes
◆ No nucleus	◆ Nucleus separated from cell by nuclear envelope
◆ No membrane-bound organelles	◆ Cytoplasm filled by membrane-bound organelles/named example of an organelle
◆ Ribosomes small (70 S)	◆ Ribosomes large (80 S)

7 3: Fungi, Plantae and Animalia.

CHAPTER 4

1 a) Abiotic factors are chemical or physical features of the environment whereas biotic factors are biological features of the environment.
b) Community refers to the populations living in an environment whereas ecosystem refers to the populations and to the abiotic factors around them.

2 a) The number of populations (different species) is large.
b) The number of individuals in each population is large.

3 Conditions on the exposed shore are much harsher for seaweeds than those on the sheltered shore. Heavy pounding by waves reduces the ability of seaweeds to grow and reproduce.

4 a) The plants help to break down the rock surface (weathering) and their dead bodies decay to produce a soil-like medium.
b) The plants are better competitors than those of the previous sere, which are unable to compete successfully, e.g. for space, water, light or nutrients.

5 a) A conspicuous mark might make the animals more easily seen by predators. This might result in our marked animals being lost from the sampling area.
b) If we do not allow time for the marked animals to disperse into the natural population, we will be more likely to capture them in the second sample and so under-estimate the true population size.

CHAPTER 5

1 A molecule of glucose breaks down too slowly and releases more energy than the cell can use. Cells can link the breakdown of ATP to anabolic reactions because ATP breaks down rapidly and releases an amount of energy that is roughly the same as that needed to drive an energy-consuming, anabolic reaction.

2 a) Water becomes part of the cytoplasm of cells; can be used for other metabolic processes (e.g. hydrolysis reactions); or is lost.
b) Carbon dioxide can be used for photosynthesis in plant cells; otherwise, it is excreted.

3 reduced NAD + carrier 1 + ADP + P_i → NAD + reduced carrier 1 + ATP

4 A compound is oxidised when it loses electrons and reduced when it gains electrons.

5 Aerobic respiration releases more energy, allowing a rapid growth in the yeast population in the early stages. Anaerobic respiration produces ethanol, which is the desired product in brewing.

6 a) RQ = volume of carbon dioxide produced/volume of oxygen used per unit time.
b) An RQ value of 0.95 suggests that the respiratory substrate is a mixture of carbohydrates and proteins.

7 reduced NADP; ATP

8 a) glycerate 3-phosphate and triose phosphate
b) chromatography

9 a) Rabbit
b) Energy is lost in the transfer from one trophic level to the next. Less energy will support a smaller biomass at each of the higher trophic levels.

CHAPTER 6

1 a) Method of nutrition where the organism is able to make organic molecules (such as carbohydrates, lipids and proteins) when provided with inorganic molecules (such as carbon dioxide and water).
b) Method of nutrition where the organism must be provided with organic molecules.

2 Low rate of photosynthesis, so small volumes of oxygen are produced. The oxygen made by photosynthesis is used by the plant for respiration.
3 Energy released from the chemical reaction is used to synthesise organic chemicals – chemosynthesis. This is a form of autotrophic nutrition.

CHAPTER 7

1 Epidermal cells take up ions from the soil solution by active transport. They also contain proteins and sugars. A higher concentration of ions, proteins and sugars makes the water potential of the cytoplasm lower (more negative) than that of the soil solution. As a result, water moves by osmosis from the soil solution into the epidermal cell.
2 The plasmodesmata are gaps in the cell walls of adjacent cells. Cytoplasm runs through these gaps from one cell to the next. These gaps increase the rate of water movement through the symplast route by removing the need for water to move through adjacent walls and surface membranes.
3 Endodermal cells actively secrete these ions into the xylem.
4 The walls of the xylem are extremely strong. Their arrangement around the outside of the stem is similar to the arrangement of steel rods in reinforced concrete, and helps to prevent the stem breaking as it is moved from side to side by gusts of wind.
5 a) cohesion
b) tension
6 Air spaces in their xylem prevent cohesion between water molecules on either side of the air spaces. Without this cohesion, water below the air space cannot be pulled up to the leaves.
7 They lose water by evaporation into the leaf's air spaces.
8 Cutting under water prevents an air bubble forming in the xylem of the cut shoot. An air bubble would prevent the cohesion between water molecules that enables water to be pulled up the stem as a result of transpiration.

CHAPTER 8

1 Essential amino acids need to be provided in the diet; non-essential amino acids can be made by transamination.
2 Water
3 Disadvantage - Need a lot of water to dilute it so that it is safe to excrete
Advantage – Saves energy necessary to convert it into something else. eg. urea
4 Urea – a nitrogen-containing compound, formed in the liver when excess protein is broken down
Urine – a solution of urea, and other salts formed in the kidney when blood is filtered
5 a) All the glucose and amino acids are reabsorbed and this must occur against the concentration gradient. The process is active therefore energy is needed. Mitochondria are the site of aerobic respiration – produce lots of ATP
b) Microvilli increase the surface area of the cell to speed up diffusion and to house more carrier molecules for active transport

CHAPTER 9

1 Excess glucose is stored as glycogen in muscle and liver. Further excess is converted into fat.
2 Cold drinks, cold shower, remove clothes, move to cooler place or into the shade.

CHAPTER 10

1 a) 6 : 1; 3 : 1; 2 : 1
b) Advantage: As very small organisms have a large surface area compared to their volume, they have no need for a specialised gas exchange surface.
Disadvantage: The large surface will tend to lose water very easily; and therefore these organisms are restricted to life in watery environments.
2 During the night/dark periods. As they are not photosynthesising, they do not produce any oxygen as a waste product, and must obtain the oxygen from the air. This travels into the plant in the same way as carbon dioxide enters during the day for photosynthesis.

CHAPTER 11

1 Carbon dioxide is soluble and, when dissolving in water, produces carbonic acid. pH falls and this affects the functioning of enzymes.
2 a) $98 - 35 = 63\%$
b) The more active the tissue, the more energy in the form of ATP is required. ATP is produced more efficiently by aerobic respiration, which needs more oxygen. The lower levels of oxygen in active tissue stimulate the greater release of oxygen from haemoglobin.

3 a) Chemical which absorbs or donates hydrogen ions.
b) Carbon dioxide forms carbonic acid when it reacts with water in the blood and therefore increases the hydrogen ion concentration. As cells increase their activity; they produce more carbon dioxide as a waste product of respiration. Without a buffer, the pH of the blood would vary with the activity of the body's cells. Enzymes in the body work at an optimum pH: changes of pH will alter the enzymes' activity.

CHAPTER 12

1 a) Glycogen
b) Liver and muscles
2 Hydrolysis
3 Triglyceride is made of three fatty acid molecules attached to glycerol; one of the fatty acid molecules is replaced by phosphate in a phospholipid.
4 Starch is made of many alpha glucose molecules; whilst cellulose is made up of many beta glucose molecules.
5 Receptor proteins on the membrane of some cells respond to the hormone. Cells which do not have that receptor, do not respond.

CHAPTER 13

1 Sensory neurone, carrying impulses from a sense organ to the central nervous system. Relay neurone, linking a sensory and a motor neurone.
2 Difference in concentration; surface area; length of diffusion pathway; temperature.
3 Chemical; physical; electrical.
4 Block vesicle release; inactivate/inhibit acetylcholine esterase; block post-synoptic receptor proteins.
5 a) Thermal/heat; kinetic/sound; solar/light; chemical.
b) Skin receptors and hypothalamus; ear; eye; tongue / nose.
6 Hot; cold; touch.
7 Light would not be focused until after it had passed through the retina. In an older eye, the ciliary muscles are not so strong; the lens remains 'thin' and light is not refracted (bent) enough.
8 Rod cells are found at the edges of the eye and they are sensitive to low light intensities – they register dim light. When you look straight at an object, the eye focuses the light on the fovea. This area of the retina contains only cone cells, which need high light intensities to react. There is not a high enough intensity of light coming from the star for the cone cells to produce a generator potential.

ANSWERS TO EXAMINATION QUESTIONS

CHAPTER 1

1 a) Anaphase I;
Chromosomes/chromatid pairs/bivalents are separating 2
(*Allow: 'they' are separating*)
b) 8; 1
c) 2; 1
d) So fertilisation/described can restore (diploid) number/prevent chromosome doubling at fertilisation/described; 1
(*Ignore references to 'variation'.*)
Total marks = 5

2 a) Gg/suitable equivalent;
Grey black about 3 : 1; 2
(*Note: Can be in table/diagram*)
b) (i) to determine the probability;
(*Accept: Likelihood*)
Of the results being due to chance:
(*Accept: Coincidence*) 2
(ii)

O	E	O–E	(O–E)^2	$\frac{(O-E)^2}{E}$
152	150]	2	4	0.027]
48	50];	2	4	0.08]

(method ignore calculation errors);
(*Note: Alternative showing of E and method*)

$$\left[\frac{(152-150)^2}{150} + \frac{(48-50)^2}{50}\right]$$

$\chi^2 = 0.107/0.11$; 3
(iii) df = 1 and $p = 0.05$/95% level *or* critical value/described = 3.84:
(*Accept: Ringed/indicated on table*)
Accept hypothesis because χ^2 is less than (table/critical) value/there is no significant difference/difference is due to chance; (*Note: Check carry forward of χ^2 value or citical value for interpretation or converse argument*) 2
c) (i) both alleles will be expressed (in the phenotype) 1
(ii) 0.25/25%;
$C^N = 250/1000$; 2
(iii) $p^2 = (0.25)^2/0.0625$/square of calculated figure for $\mathbf{C^N}$;
$p^2 + 2pq + q^2 = 1.0$;
= 31.25/31 3
(*Accept: Derived from either* p^2 *or* q^2)
Total marks = 15

CHAPTER 2

1 a) (i) Discrete groups/types/categories/explained e.g. large and small seed diameters/types exist;
(*Reject: Bimodal*) 1
(ii) Different survival advantages/explained e.g. size linked to location; selection against intermediate forms/in favour of extreme forms; 2
b) Interbreed/cross the two types of flax plants;
Offspring fertile (if same species) offspring can also interbreed/or reasonable alternative;
(*Reject: Viable*) 2
Total marks = 5

CHAPTER 3

1 a) (sympatric) organisms cannot interbreed/any valid named example, e.g., wrong courtship behaviour; 1
b) populations separated by physical barrier/named example of physical barrier;
no mixing of gene pools;
(populations) become adapated to local environment;
by natural selection;
(separated populations) accumulate different mutant alleles/diverge genetically;
allele frequencies change rapidly in small populations;
eventually the two population cannot interbreed (even if together). 4
Total marks = 5

CHAPTER 4

1 a) (i) all the organisms/populations present in a habitat;
(ii) all the individuals of one species in a habitat;
(ii) habitat and the community; 3
b) both require same resource(s)/competition for same resource(s);
named example of resource, e.g. food/light/water/ions/nest sites;
one species better adapted/competes better/survives/displaces the other; 2
Total marks = 5

CHAPTER 5

1 a) (i) CO_2 combines with ribulose bisphosphate/RuBP; (product) splits in two/production of two molecules of GP. 2
(ii) Amount formed = amount broken down/used 1

b) Any **three** from:
No ATP made (in dark);
No reduced NADP (in dark);
GP not converted (in dark);
TP not formed (in dark). 3 max

c) (i) Photosynthesis occurs during day/light; Photosynthesis uses/takes in CO_2. 2
(ii) Higher;
Less light/cooler/fewer leaves/CO_2 from soil organisms/CO_2 from decay/CO_2 from respiration in soil. 2

d) Wind mixes with air/removes CO_2/supplies more CO_2;
Introduces another variable/makes data unreliable/takes account of wind. 2

e) Any **three** from:
Respiration;
(respiration) releases CO_2;
Decomposers/microorganisms/bacteria/fungi/detritivores/worms/woodlice;
Digestion/hydrolysis (of organic matter of leaves)/decay/decomposition/rotting. 3

Total marks = 15

CHAPTER 6

1 a) 'Slash'/cutting down trees reduces photosynthesis;
Reduces removal of carbon dioxide from atmosphere;
'Burn'/combustion releases carbon dioxide.
OR 'Slash'/cutting down trees removes respiring organisms;
Reduces removal of carbon dioxide into atmosphere;
'Burn'/combustion releases carbon dioxide. 2 max

b) (Before clearing) soil exists/already produced;
(After clearing) recolonisation by new plants/seeds;
(Bring about) change in environment/soil;
(Allows) succession;
(Leading to) climax (community). 3 max

c) 1 Ammonium compounds from proteins/amino acids/urea/N-containing;
2 Converted into nitrite;
3 Into nitrate; (*Reject: Incorrect sequence once*)
4 By nitrifying bacteria/correctly named;
5 Nitrogen-fixing bacteria;
6 Fix nitrogen from atmosphere/air;
7 Nitrate taken up by plants;
8 Nitrogen needed for protein synthesis/plant growth. 6 max

d) Trees available as a sustainable resource;
Maintain habitats/niches/shelter;
Maintain diversity/avoid loss of species/protect endangered species;
Maintain stability (of ecosystem);
Maintain food chains/webs/supply of food;
Reduced loss of soil erosion; Reduced flooding;
Act as carbon sink/maintain O_2 and CO_2 balance/reduce greenhouse effect/reduce global warming; Source of medicines.
(*Ignore eutrophication*) 4 max

Total marks = 15

CHAPTER 7

1 a) (i) osmosis; 1
(ii) apoplast (pathway). 1

b) Casparian strip/waterproof walls;
so water must go through cytoplasm/vacuoles/symplast pathway. 2

c) in xylem;
evaporation/transpiration from leaves;
through stomata;
cohesion of water molecules/water molecules form hydrogen bonds;
leaf cells have more negative/lower water potential;
water drawn up as a column;
adhesion of water to walls (of xylem vessels)
capillarity due to narrow diameter/lumen of xylem (vessels);
lignified walls keep xylem (vessels) open;
root pressure forces (some) water up. 6

d) description – rise and fall, with maximum rate at midday;
rise related to increasing temperature;
fall related to stomatal closure;
explanation in terms of evaporation;
explanation of factor affecting stomatal opening/closure. 4 max

Total marks = 14

CHAPTER 8

1 a) Pituitary; 1
(*Ignore any reference to lobe/hypothalamus.*)
b) (i) (Each) protein has a tertiary structure;
Gives specific/correct shape/size to (inside of) channel/pore; 2
(ii) More negative/lower WP (inside tubule cells);
(*accept* Ψ *symbol/down a WP gradient*)
Water enters/moves by diffusion/osmosis;
(*Ignore water concentraton, etc.*) 2

Total marks = 5

2 a) (i) 'Decreases' in any upper box **and** 'increases' in any lower box. 1
(ii) Example from flowchart showing that change in water potential brings about corrective response (e.g. decrease in plasma Ψ leads to increased ADH secretion/permeability/reabsorption. 1
b) **Three** marks for **three** of:
correct reference to counter-current multiplier system;
correct differential permeability to water of the two limbs;
active transport of ions out of ascending limb/movement of water out of descending limb;
water potential gradient set up between filtrate in collecting duct and surrounding tissue;
along the length of collecting duct;
water withdrawn from collecting duct;
correct ref. to osmosis (once only). 3 max

Total marks = 5

CHAPTER 9

1 a) (i) Maintains/allows efficient/high level of activity/movement; 1
(*Ignore: Remain active.*)
OR Allows/maintains high/efficient level of enzyme reactions; 1
(*Ignore: Reactions still occur.*)
(ii) Requires more/high amount of energy/food/respiration rate; 1
(*Ignore: Loss of energy/heat.*)
b) (i) Evaporation of sweat removes heat from skin;
High(er) rate of sweating leads to low(er) skin temperature; 2
(*Ignore: Description only and Vasodilation references.*)
(ii) Change/fall in body/core temperature results in reduced sweating;
(*Reject: Stops sweating.*)
Reduced sweating results in increase in body core temperature/body core temperature returns to original level;
(*Ignore: Hypothalamus and receptors references.*)
(This) results in subsequent increase/return to original level of sweating; 2 max
(*Ignore: Description only.*)

Total marks = 6

CHAPTER 10

1 a) Accurate description of ventilation by water flow;
(Oxygen) removal by bloodstream;
Description of/countercurrent flow of blood and water (at gills);
(*Accept labelled diagram, ignore 'contraflow', reject 'multiplier'.*) 2
b) 4.0 seconds/s; (*Accept* $2 \times 2s$) 1
(Total) time when oxygen (concentration) was increasing/oxygen difusing in;
OR
(Total) time when carbon dioxide (concentation) was decreasing/carbon dioxide diffusing out. 1 max

Total marks = 4

CHAPTER 11

1 a) Picks up oxygen more readily (in lungs)/greater affinity/idea of more readily saturated;
Where O_2 is low; 2
(*Ignore 'rate of loading/unloading'.*)
b) (i) Can bind to/release H^+ ions; 1
(*Allow correct equation.*)
(ii) Denature/alters shape/charge of active site;
Breaking/forming bonds/named bond;
reject 'peptide';
Substrate 'repelled'/cannot fit/no E–S complex. 2 max

Total marks = 5

CHAPTER 12

1 a) Larva has varied diet/described (to supply materials and energy) for growth;
Larva needs a range of enzymes to digest all nutrients in diet;
Adult only eats nectar/sucrose as no growth/only reproduction in adult phase or adult only eats nectar/sucrose as only needs energy for movement;
Adult only needs sucrase to hydrolyse sucrose (in nectar). 3 max

b) Carbohydrate stored in leaves as starch;
Hydrolysed to glucose/converted by amylase to glucose. 2

Total marks = 5

CHAPTER 13

1 a) (Pressure), deforms/opens (sodium) channels;
(*Reject any other ion.*)
Sodium ions enter;
Causing depolarisation;
Increased pressure opens more channels/greater sodium entry. 2 max

b) (i) Arrow (labelled K) pointing out of node; 1
(ii) Same amplitude of action potentials as in medium pressure graph but of a greater frequency. 1

c) (i) Answer between 0.7 and 0.9 (ms). 1
(ii) Correct answer based on candidate's response to c) (i) (i.e. 80 divided by answer to previous question) 1
(*Accept correct working shown with no final answer.*)

d) (i) Action potential/impulse unable to 'jump' from node to node/no saltatory conduction/action pd/ impulse must pass through a greater amount of membrane;
Slows/prevents impulse. 2 max
(ii) Greater entry of sodium ions/greater exit of K^+ in de-myelinated neurone;
Ref. to active transport/ref. to ion pumps. 2

Total marks = 10